Lecture Notes in Electrical Engineering

Volume 933

The book series *Lecture Notes in Electrical Engineering* (LNEE) publishes the latest developments in Electrical Engineering - quickly, informally and in high quality. While original research reported in proceedings and monographs has traditionally formed the core of LNEE, we also encourage authors to submit books devoted to supporting student education and professional training in the various fields and applications areas of electrical engineering. The series cover classical and emerging topics concerning:

- Communication Engineering, Information Theory and Networks
- Electronics Engineering and Microelectronics
- Signal, Image and Speech Processing
- Wireless and Mobile Communication
- Circuits and Systems
- Energy Systems, Power Electronics and Electrical Machines
- Electro-optical Engineering
- Instrumentation Engineering
- Avionics Engineering
- Control Systems
- Internet-of-Things and Cybersecurity
- Biomedical Devices, MEMS and NEMS

For general information about this book series, comments or suggestions, please contact leontina.dicecco@springer.com.

To submit a proposal or request further information, please contact the Publishing Editor in your country:

China

Jasmine Dou, Editor (jasmine.dou@springer.com)

India, Japan, Rest of Asia

Swati Meherishi, Editorial Director (Swati.Meherishi@springer.com)

Southeast Asia, Australia, New Zealand

Ramesh Nath Premnath, Editor (ramesh.premnath@springernature.com)

USA, Canada:

Michael Luby, Senior Editor (michael.luby@springer.com)

All other Countries:

Leontina Di Cecco, Senior Editor (leontina.dicecco@springer.com)

**** This series is indexed by EI Compendex and Scopus databases. ****

More information about this series at https://link.springer.com/bookseries/7818

Miguel Botto-Tobar · Marcelo Zambrano Vizuete ·
Angela Diaz Cadena · Ana Zambrano Vizuete
Editors

Latest Advances in Electrical Engineering, and Electronics

Proceedings of the JIEE 2021

Springer

Editors
Miguel Botto-Tobar
Eindhoven University of Technology
Eindhoven, The Netherlands

Marcelo Zambrano Vizuete
Universidad Técnica del Norte
Ibarra, Ecuador

Angela Diaz Cadena
Universitat de Valencia
Valencia, Spain

Ana Zambrano Vizuete
Escuela Politécnica Nacional
Quito, Ecuador

ISSN 1876-1100 ISSN 1876-1119 (electronic)
Lecture Notes in Electrical Engineering
ISBN 978-3-031-08941-1 ISBN 978-3-031-08942-8 (eBook)
https://doi.org/10.1007/978-3-031-08942-8

This Springer imprint is published by the registered company Springer Nature Switzerland AG
The registered company address is: Gewerbestrasse 11, 6330 Cham, Switzerland

Contents

Electric Power

Electric Energy Demand Forecasting in an Oil Production Company Using Artificial Neural Networks

Alex Manobanda, Patricia Otero(✉), and Nelson Granda

Electric Energy Department, Escuela Politécnica Nacional, Quito, Ecuador
{alex.manobanda,patricia.otero,nelson.granda}@epn.edu.ec

Abstract. Electric energy demand forecasting is vital for the correct and efficient operation of any industry, especially for those ones that need to purchase energy. Activities such as maintenance, new installations, and energy purchase, are planned based on the demand forecasting. Therefore, it is necessary to use an advanced technique to obtain a forecasting as accurate as possible. In the present paper, a methodology based on artificial neural networks is developed for forecasting the electric energy demand in an oil production company in Ecuador, based on fluid production and historical data. The forecasting technique was applied using Python programming language. Finally, the results are compared with the conventional simple linear regression methodology, proving a satisfactory precision.

Keywords: Electricity demand · Demand forecasting · Artificial neural networks · Multilayer perceptron neural network

1 Introduction

A magnitude forecasting consists in advanced calculations that are made of an event that is going to happen, based on some determinate signals or indications [1]. To make the best technical and economic decisions, industries predict the future behavior of the variables involved in the operations.

There are a variety of methodologies to predict electricity demand, among which are the traditional methods that use statistics as a basis such as: time series, regressions, and econometric models. In addition in recent years more advanced techniques have been developed (artificial intelligence) such as: artificial neural networks (ANN), genetic algorithms, fuzzy logic, among others [2, 3]. The different methods have their own technical, mathematical, and computational characteristics, with advantages and disadvantages that depend on the application in which they are used and the type of data with which they work.

Due to the technological advance, essentially in the computational processing, more complex algorithms have been implemented such as the ANN that bring with them more reliable results in the forecasting, these algorithms are able to learn trends and behaviors

M. Botto-Tobar et al. (Eds.): *Latest Advances in Electrical Engineering, and Electronics*, LNEE 933, pp. 3–16, 2022.
https://doi.org/10.1007/978-3-031-08942-8_1

from the historical data and thus can perform better forecasting avoiding overestimation or underestimation of electricity demand [4, 5].

This article is organized in the following way. In Sect. 2, the state of the art regarding the ANN is discussed. In Sect. 3, the proposed electricity demand is presented. In Sect. 4, a case study is made with data from an oil production company (OPC), the results obtained in the proposed methodology and the current methodology are showed and compared. Finally, the conclusions are presented in Sect. 5.

2 Artificial Neural Networks (ANN)

The ANNs are models inspired by the nervous system of the human being, which consists, in the interconnection of neurons that allow to process the information and obtain the desired output according to the needs of the human being. ANNs are structures composed of interconnected artificial neurons, which implement computational functions to obtain the desired outputs [6, 7]. They allow to model non-linear behaviors, being very useful for the electricity demand forecasting in an OPC that presents this characteristic.

The multilayer perceptron is the ANN that has presented the best results for this particular application, it is composed of an input layer, one or more hidden layers and an output layer (see Fig. 1) [8, 9], this architecture has the following characteristics:

- There are no connections between neurons in the same layer.
- The activation functions are the same for each neuron in the same layer.
- It has a single neuron in the output layer.

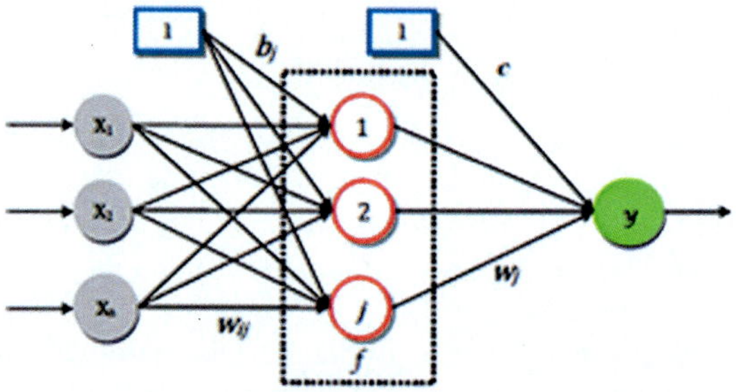

Fig. 1. Multilayer perceptron architecture.

The mathematical representation of a multilayer perceptron is given by the following equation:

$$y = \sum_{j=1}^{m} w_j f_j \left(\sum_{i=1}^{n} a_{ij} x_i + b_j \right) + c \tag{1}$$

where y is the estimated variable, x_i are input variables, w_j are the connection weights between the *j-th* neuron of the last hidden layer and the output layer, w_{ij} are the connection weights between the *i-th* input to the model with the *j-th* neuron of the first hidden layer, b_j are the bias of the *j-th* hidden neuron, *c* represents the bias of the output neuron.

w_j, w_{ij}, b_j and *c* are the parameters of the model that will be adjusted through a certain training algorithm. From the equation can be extracted that the model represents a regression characterized by the function f (linear or nonlinear) that determines how the inputs and outputs of the model are related [10]. The activation functions that allow to model the performance of the electricity demand (see Fig. 2) are: hyperbolic and linear tangent.

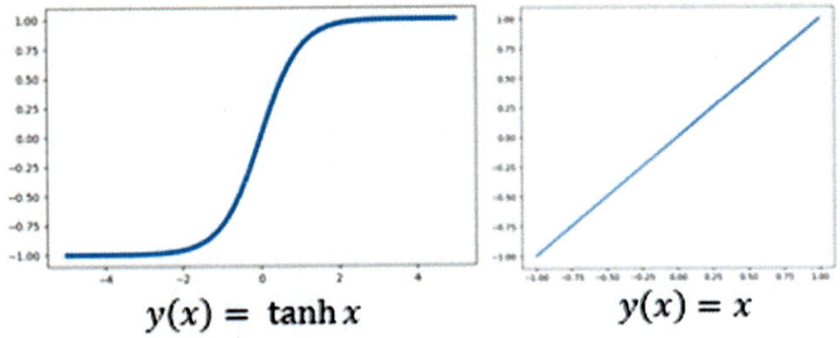

Fig. 2. Activation functions of an ANN.

3 Methods

The following steps represents the basic methodology for the electricity demand forecasting using ANN:

- **Step 1.** Understanding the operation of the company is related to searching for all the information that allows understand the business (production and historical data). This allows to interpret the database that the company has.
- **Step 2.** Select input variables according to the application and horizon to be predicted. Based on the knowledge acquired in step 1, it is necessary to choose the variables that most influence the behavior of the electricity demand.
- **Step 3.** Pre-processing of input data. The ANN learns and predicts based on the historical data of the variables considered, if in these databases there are erroneous data, the results of the forecasting will not be the most accurate, that is why prior to the insertion of the bases to the model, it is necessary to eliminate the erroneous data that may exist.
- **Step 4.** Define the ANN (number of input neurons, hidden layers, neurons in the hidden layers, output neurons and activation functions). The most important step to do, often depends on the experience of the predictor, it is necessary to compare between different ANN structures of ANN to find the one that best suits the application.
- **Step 5.** The ANN training (learning criteria, number of iterations). In this step, the final values of the weights between the different neurons and the bias of the ANN are assigned, minimizing the value of the selected learning criterion.

- **Step 6.** Verification of the ANN. It is performed by predicting values of electricity demand and comparing them with real data.
- **Step 7.** Storage of the ANN parameters. In accordance with the criteria of the company, the ANN is ready for future use, without the need to repeat the last two steps unless those responsible believe it necessary.
- **Step 8.** Implementation of the ANN for the forecasting of the electricity demand.

The steps are described below.

3.1 Prior Knowledge of the Company

The OPC is an Ecuadorian company, whose name will not be disclosed for confidentiality reasons. The company operates in the Ecuadorian Amazon Region and is dedicated to the exploration and production of hydrocarbons.

The OPC oversees the operation of its fields, most of them located geographically isolated from the national interconnected system of electricity, therefore the planning includes the amount of energy that is needed. Each facility has: production wells, equipment for the extraction, treatment, and storage of hydrocarbons. The production process is mentioned below.

Fluid Extraction. Stage in which the fluid is extracted from the reservoirs to the surface, using the following techniques: natural fluid that uses the intrinsic energy that exists in the reservoir and artificial lifting that uses external mechanisms (mechanical, hydraulic or electro submersible pumping and gas lift).

Fluid Processing. At this stage, the external fluid is separated into its basic components (oil, water, and gas) and all the undesirable components found in it are discarded.

Fluids Storage. The production process ends with the storage of the components obtained: the oil is stored for post-distribution; the gas is used in electricity generation or is expelled into the environment and the water is reinjected into the oil wells.

Water Reinjection. It allows to improve or increase the time of production of the wells, since by means of pumps water is sent to the deposits.

On the other hand, for the information acquisition and storage the does the following:

- Fluid production records are determined by static measurement or dynamic measurement in each field, these records are measured in fluid barrels per day (FBPD) and stored in Excel format.
- The electrical demand records are obtained from meters installed in the fields, these records are measured in kWh and stored in Excel format.

3.2 Selection and Analysis of the Input Variables Historical Data

The production of an industry is the most influential factor in the behavior of electricity demand, because for a higher production the greater electricity demand is. As the OPC

Table 1. Variable autocorrelation.

	Crude oil	Water	Gas	Fluid	Electricity demand
Crude oil	1	0,9498	0,8496	0,9670	0,9062
Water	0,9498	1	0,8376	0,9982	0,9123
Gas	0,8496	0,8376	1	0,8465	0,7977
Fluid	0,9670	0,9982	0,8465	1	0,9183
Electricity demand	0,9062	0,9123	0,7977	0,9183	1

has different extraction products (crude oil, gas, and water), an autocorrelation analysis is developed for a correct selection of the input variables for the model.

Table 1 shows all the variables are highly autocorrelated, ergo, they all have a similar behavior with respect to the electricity demand. It is selected the variable that has the most correlation with the electricity, the fluid (the total of crude oil plus water).

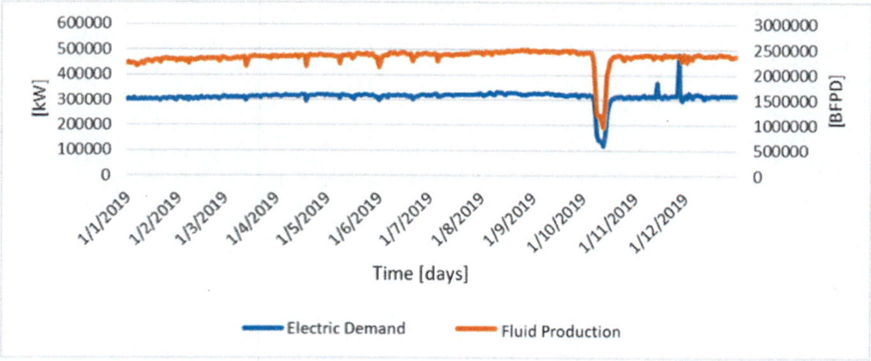

Fig. 3. Total electrical energy demand and total fluid production over time.

Figure 3 presents the electricity demand and the fluid behavior in time and Fig. 4 shows the relation between the two variables mentioned. From these figures, it is evident that if the fluid production increases the electricity demand too, but this growth is not entirely linear, due to the differences found in the operation of each field.

The relatively small fields cannot be compares with the large ones. On the other hand, the efficiency of the assets is not the same in all of them, there will be some fields that need greater demand for electrical energy to produce the same amount of fluid, this effect is reflected when comparing the average data of energy efficiency index (EEI) of the different assets. The field with a smaller EEI value is more efficient, these differences are summarized in Table 2.

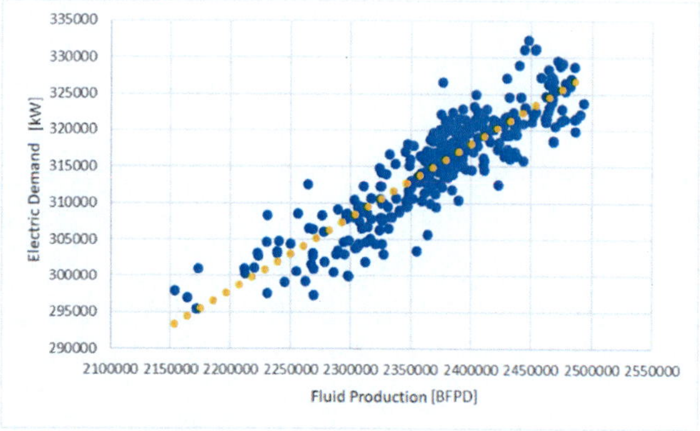

Fig. 4. Electricity demand vs total fluid production.

Table 2. Fields summary

Field	Average production [BFPD]	EEI
Field 1	89000	0.06553
Field 2	170000	0.19835
Field 3	123000	0.12909
Field 4	380000	0.13804
Field 5	198000	0.14493
Field 6	17000	0.43416
Field 7	131000	0.15722
Field 8	250000	0.11307
Field 9	70000	0.12685
Field 10	155000	0.18966
Field 11	306000	0.14799
Field 12	500000	0.07457

3.3 Data Pre-processing

The OPC has a relatively good database, however, as in any database, it is possible to find atypical and missing values due to a variety of reason like unexpected events or communication failure.

For the OPC studied, an unexpected event occurred in October 2019, a national strike that affected the oil production, generating atypical data. Additionally, there are atypicals of unknown origin, these values must be eliminated to avoid errors in training the ANN. The elimination is done by the software operator, based on his experience and knowledge of the events [11].

Another case is the inexistent data, the missing value is substituted or replaced by the average of the variable that is calculated with the available data [11, 12].

3.4 ANN Implemented

The type of ANN to be implemented is the multilayer perceptron, which has brought satisfactory results for forecasting. The parameters selection depends on the experience of the operator. Usually, different ANN configurations are compared and the one with best results is selected, as it is shown in [13–15].

Selection of the Number of Layers. ANNs can be composed of an input layer, an output layer, and one or more hidden layers. The number of hidden layers is the first parameter that will be compared. Generally, with the smallest number of hidden layers it is usually obtained great results for forecasting. For the present application the following configurations are examined with respect to the number of hidden layers: first an ANN that presents 1 hidden layer and second an ANN that presents 2 hidden layers. The configuration of the number of hidden layers together with the configuration of the number of neurons of the ANN are the two fundamental parameters that differentiate the proposed cases studied [15].

Selection of the Number of Neurons. The number of neurons for the input layer is defined by the number of input variables set, therefore, having 12 inputs (production of barrels of oil per asset), the number of input neurons has been defined at 12. The number of neurons for the output layer is defined by the number of output variables set, therefore, having 1 output (electrical energy demand), the number of output neurons has been defined at 1 [15].

There is no rule for the assignment of neurons to each hidden layer, it is necessary to test a different number of neurons in the ANN and select the one that improves the accuracy of the forecasting. The number of neurons per layer is another parameter that will be compared. For the present application, the cases that can occur in the definition of the number of neurons will be studied:

- The number of neurons in the layers increases (relative to the number of input neurons) as it travels through the network.
- The number of neurons in the layers decreases (relative to the number of input neurons) as it travels through the network.
- The number of neurons in the layers remains constant (relative to the number of input neurons) as it travels the network.

Activation Function Selection. The sigmoidal function or hyperbolic tangent are used in the hidden layers because they give the ANN the ability to learn possible nonlinear behaviors. By using these activation functions the ANN is robust even to atypical values that may be present. For the present case of application, it is chosen to use as an activation function the hyperbolic tangent for the hidden layers, since the data normalization is in the range of -1 to 1. And the linear activation function for the output layer.

3.5 Study Cases

Seven cases were created to compare different ANN configurations in the forecasting of electricity demand, with the objective to select the one that provides the best results. The first 6 cases present as input variables the fluid production per field (12 input variables), and the last case presents as input variable the total fluid production (1 input variable). Table 3 details the cases compared for this case.

Table 3. Study cases.

Case	Layers	Neurons number	Activation function
Case 1	Input	12	–
	Hidden 1	12	Hyperbolic tangent
	Output	1	Linear
Case 2	Input	12	–
	Hidden 1	24	Hyperbolic tangent
	Output	1	Linear
Case 3	Input	12	–
	Hidden 1	6	Hyperbolic tangent
	Output	1	Linear
Case 4	Input	12	–
	Hidden 1	12	Hyperbolic tangent
	Hidden 2	12	Linear
	Output	1	Linear
Case 5	Input	12	–
	Hidden 1	12	Hyperbolic tangent
	Hidden 2	12	Linear
	Output	1	Linear
Case 6	Input	12	–
	Hidden 1	12	Hyperbolic tangent
	Hidden 2	12	Linear
	Output	1	Linear
Case 7	Input	12	–
	Hidden 1	12	Hyperbolic tangent
	Output	1	Linear

For each case the mean square (MSE) and the mean absolute percentage error (MAPE) are calculated based on real data and projected data, these indicators will allow to compare the goodness of the different study cases, and the selection of the ANN that provides the minimum value in the indicators.

3.6 Training, Validation and Storage of the ANN

For the present case of application, it was decided to use a supervised learning mechanism. The backpropagation method is used for ANN training. Another parameter to be defined is the validation criterion that is to be minimized in training, for the present case is the mean square error.

Once the ANN is trained, the next step in the methodology is the validation of the model for that it is necessary to use the subset of test data of fluid production, which will be entered into the trained model and the forecasting of electric power demand are compared with the subset of test data of historical electrical energy demand, once again the mean square error will be used as a validation criterion.

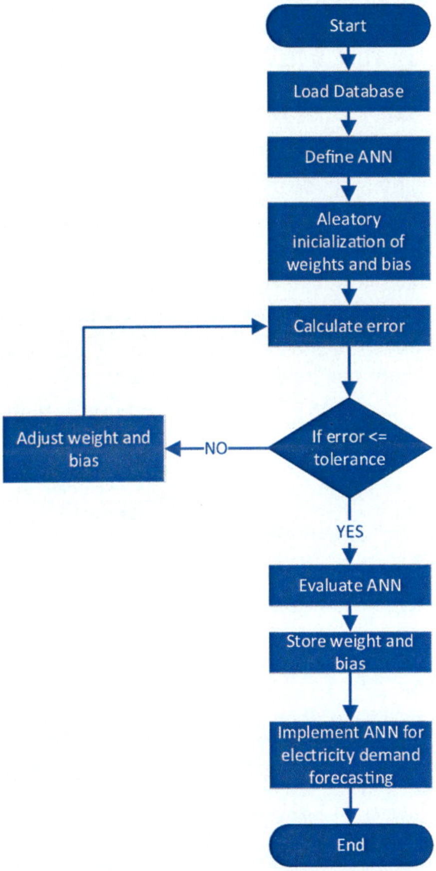

Fig. 5. Training, validation and storage flowchart

The model that presents the lowest value in the mean square error will be stored (for later use in the energy planning. Figure 5 shows the process flowchart for the training, validation, and storage of the ANN.

3.7 Electricity Demand Forecasting Using Python

The Python programming language was selected because it uses high-level structures that allow a simple but effective approach to object-oriented programming. Also, its characteristics for the syntax and dynamic typing, make it an ideal language for scripting and for the development of applications in different areas (artificial intelligence, big data, data science, web development, among others) [16].

The libraries used for this application are [16]:

- Numpy. Is the main Python package for scientific computing. Provides data structures (implementing multidimensional matrices and matrices), a variety of routines for fast operations with matrices, which include mathematical operations, manipulation of forms of data structures, among others.
- Pandas. Provides fast, flexible, and expressive data structures designed to work with "labeled" data (series and dataframe) that facilitate data manipulation and analysis. It allows: to clean the data, analyze the data, and finally organize the results of the analysis in a form suitable for visualization.
- Matplotlib. Is a package used for plotting (2D graphics), it is very flexible (variety of printed formats), it has many built-in defaults that make the job easier.
- Scikit-Learn. Implements a variety of machine learning algorithms (regression, classification, grouping).
- Keras. Allows to implement high-level neural networks (convolutional networks, recurrent networks, as well as combinations of the two) that are executed on TensorFlow. It is a library to implement deep learning to programmers.
- Tensorflow. Is a package for machine learning. It presents a comprehensive and flexible ecosystem, libraries and community resources that allows applications to be developed in an easy way.

Figure 6 presents the logic that follows the Python programming of the electric power demand forecasting module, using a flowchart.

4 Results

4.1 Historical Data Analysis

The program developed in Python, allows the user to analyze the historical data providing a displaying the electricity demand and fluid production as shown in Fig. 7.

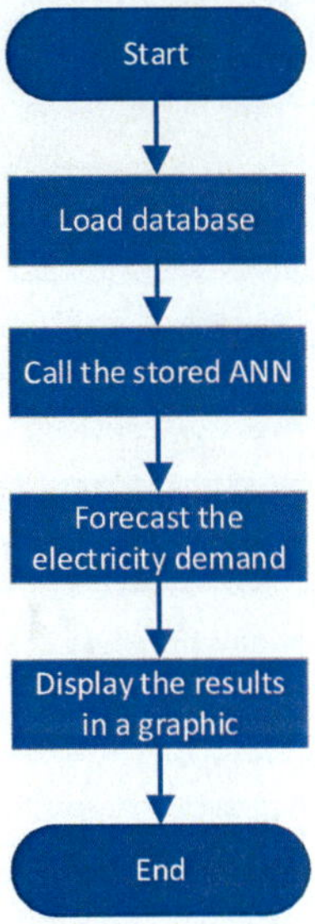

Fig. 6. Flowchart for electricity demand forecasting

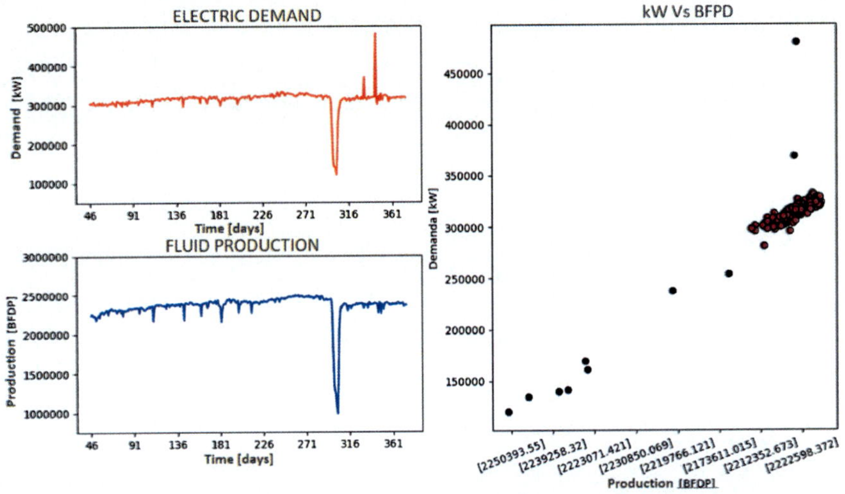

Fig. 7. Graphical interface module for analysis of historical data

4.2 ANN Configuration Selection and Training

Following the methodology, the parameters to train the proposed ANN are defined. A supervised learning mechanism is used, and the backpropagation method, the MSE and MAPE are used as a validation criterion. Once the ANN is trained, the model is validated with a subset of test data, which enters to the trained model and the results are compared with the historical data of electricity demand. The configuration that presents the least error, will be stored for later use in the energy forecasting. The results are shown in Table 4.

Table 4. ANN configuration results

Configuration	MSE	MAPE [%]
Simple linear regression	426779742.52	1.49
Case 1	13669531.22	0.86
Case 2	22545017.12	1.28
Case 3	29815201.78	1.52
Case 4	18733288.41	1.09
Case 5	23979040.58	1.31
Case 6	24765789.01	1.32
Case 7	9648639.19	0.73

Most of the proposed MAPE and MSE in the case studies are lower than that obtained by simple linear regression except for case 3. The ANN configurations that gave the best results are case 1 and case 7, which are differentiated by the number of input variables.

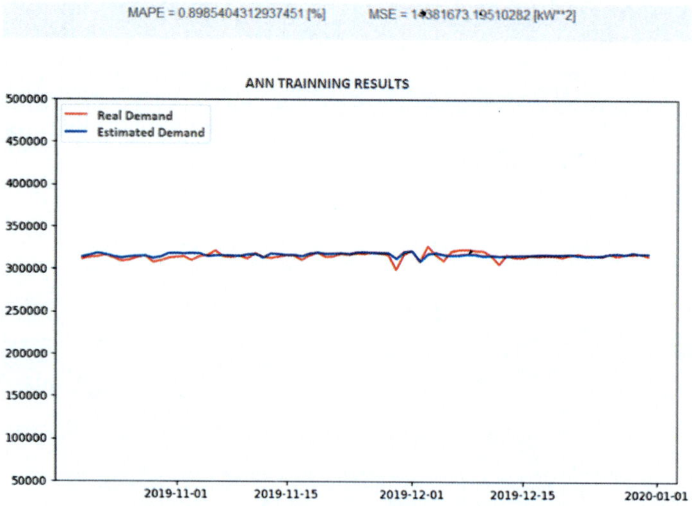

Fig. 8. Graphical interface showing the ANN training.

The case 1 configuration is selected because it contemplates changes in fluid production by asset (not all grow or decrease their production at the same time). Figure 8 shows the ANN training result for this case.

4.3 Electricity Demand Forecasting

The electricity of the OPC in 2020 tends to grow over time (see Fig. 9), at the beginning of the year it stands at a value of 318000 [kW] and at the end of the year it will stand at a value close to 325000 [kW], this is due to the fluid production throughout the year will have a gradual growth.

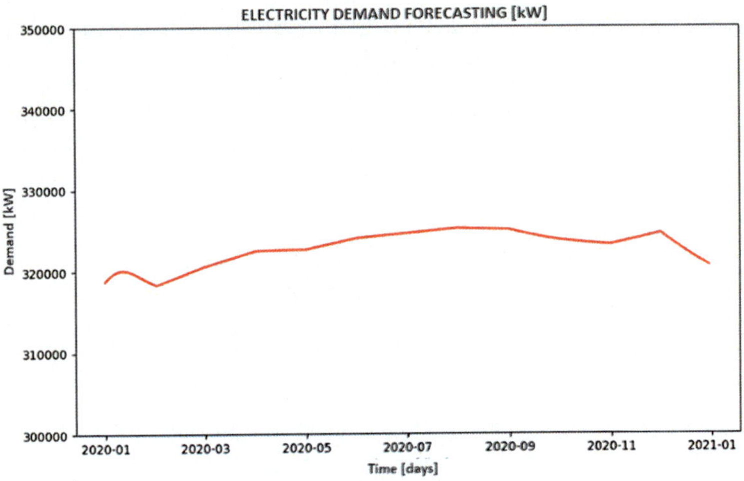

Fig. 9. Forecasting of electricity demand

5 Conclusions

ANNs are artificial intelligence algorithms that allow to improve the forecasting of the demand of electrical energy, because it allows to emulate the non-linear behavior that presents the demand of electrical energy with respect to the production of fluid, thus allowing planners to reduce uncertainty in making strategic decisions.

For the realization of the forecasting of the demand for electrical energy it is necessary to follow certain steps: first a prior knowledge of the company (company name, production and storage of information), after this study and select the different input variables, then validate the information extracted from the historical databases, to finally create, train and validate the ANN to be used.

Eliminating or imputing the missing and outlier values of a historical database allows to improve the accuracy in the results of the electric demand forecasting.

The increase in the number of hidden layers or number of neurons of the artificial neural network does not imply better results, in many cases the artificial neural network is overtrained affecting the forecasting.

It is necessary to develop the selection of the ANN configuration through trial and error until finding the configuration with the least error present in the training and testing stage.

It was determined that the ANN that best behaved before the application are those exposed in CASE 1 and CASE 7, reducing the error of the forecasts to almost half, which shows that a neural network can be used as an algorithm to predict the demand for electrical energy.

References

1. Gómez, M.: Modelo de previsión de demanda de electricidad de largo plazo. tesis de grado, Univ. Pontificia Comillas, Madrid, Esp. (2010)
2. Feinberg, E.A., Genethliou, D.: Load Forecasting – Semantic Scholar. In: Applied Mathematics for Power Systems, 1st edn, pp. 269–285. Springer, New York, USA (2005)
3. Singh, S., Hussain, S., Bazaz, M.A.: Short term load forecasting using artificial neural network." trabajo de fin de máster, Thapar Institute of Engineering & Technology, Patiala, Punjab (2018)
4. Chen, S.-F.: Short- and mid-term load forecasting using machine learning models. In: China Int. Electr. Energy Conf., no. 51507193, pp. 406–411 (2017)
5. Ghods, L., Kalantar, M.: Different methods of long-term electric load demand forecasting ; a comprehensive review. Iran. J. Electr. Electron. Eng. 7(4), 249–259. 10.170695094 (2011)
6. Bunnoon, P.: Mid-term load forecasting based on neural network algorithm: a comparison of models. Int. J. Comput. Electr. Eng. **3**(4), 600–605 (2011). https://doi.org/10.7763/ijcee.2011.v3.388
7. Hernández, E., Vera, E., Vásquez, P.: Mid-term load forecasting – application in oil facilities Eddison. IEEE Ind. Appl. Soc. Anu. Meet. **2**, 1–6 (2016)
8. Param, S.: Electricity Demand Prediction Using Artificial Neural Network. trabajo de fin de máster North Dakota State Univ, North Dakota, Fargo (2015)
9. Bunnoon, P., Chalermyanont, K., Limsakul, C.: A computing model of artificial intelligent approaches to mid-term load forecasting: a state-of-the-art- survey for the researcher. Int. J. Eng. Technol. **2**(1), 94–100 (2010). https://doi.org/10.7763/ijet.2010.v2.106
10. Pérez, A.: Aprender Inteligencia Artificial, Combinatoria, Grafos y Algoritmos en Python con 100 ejercicios prácticos, 1ra. ed. Marcombo, S.A., Barcelona, Esp. (2018)
11. Segura, E., Torres, V.: Tratamiento de valores perdidos y atípicos en la aplicación del Modelo Estadístico de Medición de Impacto en un estudio de 90 fincas lecheras en la provincia de Pastaza, Ecuador. Rev. Cuba. Cienc. Agrícola **48**(4), 333–336 (2014)
12. Flores Urgiles, C.M., Ortiz Amoroso, M.S.: Revisión de algoritmos para la detección de valores atípicos. Kill. Técnica **2**(1), 19 (2018). https://doi.org/10.26871/killkana_tecnica.v2i1.287
13. Hilera, J., Martínez, V.: Redes Neuronales Artificiales Fundamentos, modelos y aplicaciones, 1ra. Ed. DF., Mex.: Alfaomega ra-ma (2000)
14. Bertona, L.: Entrenamiento de Redes Neuronales Basado en Algoritmos Evolutivos. tesis de grado, Univ. de Buenos Aires, Buenos Aires, Arg. (2005)
15. Mehrotra, K., Mohan, C., Ranka, S.: Elements of Artificial Neural Networks, 2nd edn. MIT Press, Massachusetts, USA (2000)
16. P. Software Foundation.: The Python Tutorial. Python (2019)

Secondary Voltage Control Areas Using Hybrid Methods for Systems with High Wind Penetration

S. Vaca(✉) and C. Gallardo

Departamento de Energía Eléctrica, Facultad de Ingeniería Eléctrica y Electrónica, Escuela Politécnica Nacional, Ladrón de Guevara, E11-253 Quito, Ecuador
esteban.vaca01@epn.edu.ec

Abstract. The problem of voltage stability has become a major challenge in current power systems, where there is high penetration of wind energy and ineffective voltage control schemes to adequately take advantage of reactive power reserves. This work presents a novel proposal for secondary voltage control, which considers quasi-optimal control areas determined based on Modal Energy criteria and a MonteCarlo Simulation Approach combined with K-means clustering. The simulations consider the concept of Extended Power Flow, the inclusion of stochastic models for Wind Power and the optimization of reactive power reserves for control. This proposal is intended to optimize reactive power resources when voltage setpoints are defined in pilot area buses for control purposes. The results of the proposed control strategy are validated by Residue Method Evaluation and Quasi-Dynamic Simulation. In the determined control areas, the most critical buses are identified. In this buses, reactive compensation devices are incorporated to contribute to the voltage control of the areas. The proposed methods are applied in the IEEE 39-bus system. The obtained results clearly show the effectiveness of the proposed control strategy. In addition, a statistical comparison with results in which there is no control is presented.

Keywords: Control areas · Modal energy · MonteCarlo simulation · Quasi-Dynamic simulation · Voltage stability

1 Introduction

Power Systems eventually operate close to their limit capacity, which entails risk situations that can trigger problems such as instability and the disconnection of elements (including partial and total blackouts). One of these situations is voltage instability, produced by the inability of the system to supply the required reactive power. Voltage instability can cause decline in voltage at different points in the system and possibly a voltage collapse. For this reason, to provide a continuous and a stable operation, control strategies based on the actual operating points of the system voltages are required. Since voltage is a local variable in each bus of a system, individual control is inefficient and, in many cases, impossible considering the large number of elements in the system.

M. Botto-Tobar et al. (Eds.): *Latest Advances in Electrical Engineering, and Electronics*, LNEE 933, pp. 17–32, 2022.
https://doi.org/10.1007/978-3-031-08942-8_2

Presently available research works for Secondary Voltage Control are focused on studying control methodologies, which simplifies this problem through the division of the system into control areas. These areas are defined in such a way that they can be controlled independently of the others. Own control strategies are defined in each area. This division is achieved with the definition of pilot buses in each area and the corresponding buses related with each one.

In [1], a methodology that divides a power system into control areas, so that control over them is independent or does not have significant impact on other areas is firstly proposed. This is done by searching for representative buses in each area. This search and partition of the system is carried out with the Higher Modal Energy technique. However, only specific operation points or scenarios are considered. Several robust strategies for partitioning are developed in [2–4], which use concepts from linear algebra, Modal Analysis and Heuristic criteria to represent the relationship between the buses of a system and their participation in the secondary control. Nevertheless, the application of these concepts in more complex cases such as power systems with wind power has not yet been addressed. In [5], different methods to select pilot buses are described and compared. The global search algorithm, the extended greedy algorithm of order n, the local search and the simulated annealing algorithm are studied. In [6], a comparative study of methodologies for identifying voltage control areas is performed.

Some works perform vulnerability analysis of each determined area through the oriented search of vulnerable buses, according to several methodologies given in [7, 8]. Critical voltages in each of these buses are estimated to establish safe operating conditions through the voltages to be controlled in the corresponding pilot buses. In this sense, different voltage stability indexes are proposed in [9, 10], which provide a timely warning during a sequence of events leading to voltage collapse. However, a detailed simulation of generation or the different load levels is not available.

On the other hand, some research works, have been carried out studies for secondary control considering Data Mining or Arithmetic Methods including uncertainty and renewable energy [3, 6, 11, 12]. A methodology to implement a hierarchical voltage control based on a multilayer architecture is proposed in [13], where high wind penetration is firstly simulated. A tertiary control is considered to perform an Optimal Power Flow to determine voltage setpoints for pilot buses. In the secondary voltage control, primary control variables are adjusted to achieve the desired setpoints. The voltage profile is optimized to enhance system security. In [14], a secondary voltage control scheme is proposed considering system-wide information, and different operating points, load levels and different network topologies, while keeping computational efficiency. A probabilistic determination of pilot buses for secondary voltage control is depiçted in [15], where uncertainty of operational and topological disturbances of the power system are included. In this sense, a scenario reduction technique is used to provide robustness for the pilot bus selection. The problem is solved with the support of a new immune-based genetic algorithm. In [16, 17] new approaches for pilot bus selection in applications of secondary control voltage are developed. The physical response of generators and wind power fluctuation is considered. Finally, in [18], the author brings a selection of several research works together focused on secondary voltage control, where the concepts of Holomorphic Embedding Load Flow Method, Numerical Optimization and real implementations are discussed.

In summary, although researchers have given different viewpoints regarding pilot bus selection and secondary voltage control, there are not clearly nor completely defined practices for this analysis according to each application. In this way, the next drawbacks still need to be overcome:

- Requirements of simulations which provide more detail in the generator response including wind power and load levels.
- Consideration of extensive and representative scenarios, which consider different load levels, uncertainty in the wind power and alternative reactive power sources such as SVC elements.
- Availability of a methodology for secondary voltage control that can be applied in a general way for actual power systems.

In this project, a novel proposal for secondary voltage control is developed. The proposal considers quasi-optimal control areas partitioning based on Modal Energy criteria and a MonteCarlo Simulation Approach combined with K-means clustering. The simulations consider the concept of Extended Power Flow (EPF), the inclusion of stochastic models for Wind Power and the optimization of reactive power reserves for control. This proposal is intended to optimize reactive power resources when voltage setpoints are defined in pilot area buses for control purposes.

The remainder of this document is organized in this way. Section 2 introduces the theoretical background of the proposed methodology. Section 3 describes the Hybrid methods for control areas partitioning and Secondary Control analysis. In Sect. 4 the most important results and their analysis is exposed. Finally, in Sect. 5 the main conclusions are presented.

2 Theoretical Background

2.1 Voltage Instability Analysis

Voltage stability is defined as the ability of a power system to maintain acceptable voltages at all buses in the system, under normal condition and after being subjected to a disturbance. Voltage instability occurs when there is an inability of the system to supply the required reactive power for said condition [19]. The basic criterion of voltage stability, defined in [20], is to guarantee that, for each bus, an increase in the reactive power injected causes a rise in the voltage. If an increase in the reactive power produces a decrease in the bus voltage, the system is unstable and close to a voltage collapse.

Voltage stability depends on several factors, such as the system configuration, power transfer levels, strength of the transmission network, reactive power compensation, and their location. In the same way as in the analysis of other types of instabilities, there is a hierarchical control. In this sense, the possibility of controlling the bus voltages of an area by means of representative buses emerges [4]. Note that voltage stability analysis, in practice, is carried out with different objectives, for example, searching for optimal voltage levels, the required reactive power resources in the short, medium, and long term or stability margins [1, 21].

2.2 Voltage Control Areas and Pilot Buses

Ideally, the monitoring and control of every voltage of a system in the short, medium, and long term would make it possible to avoid problems of voltage instability, guaranteeing that the most adequate reactive power resources always exist. However, current power systems are extensive networks of great magnitude, whose decentralized monitoring of each of its components is not necessarily the most efficient and, at times, becomes an impossible task [4]. In practice, data from voltages in real time for every bus in a system is extensive and it becomes hard to use for operators of the system. This situation has motivated researchers to look for ways to concentrate voltage control techniques in one dimension of areas, i.e., partitions of the system where controls are carried out in each area independently through representative buses known as pilot buses. The identification of control areas in a power system is a process that is oriented according to the mathematical development of [1] and [4] to identify groups of buses that are sufficiently decoupled from other groups. Each voltage control area is composed only of buses that have significant electrical coupling. Each of the areas is formed around a representative or pilot bus that has observability and controllability. A schematic of a partitioned power system in control areas is shown in Fig. 1. It is desired that each of the areas have little or no influence with respect to the others.

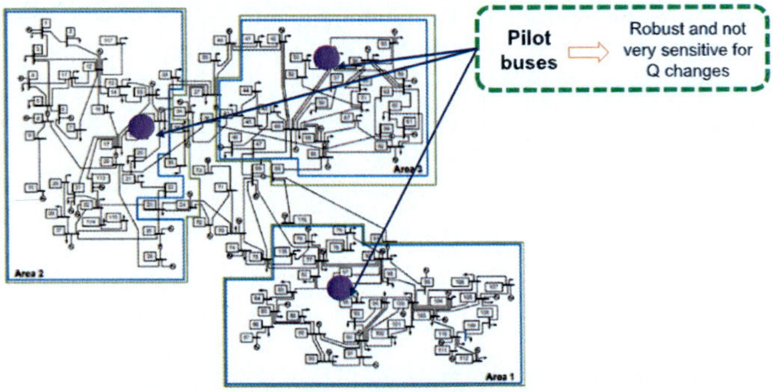

Fig. 1. Schematic of pilot buses for control areas

Some of the advantages of voltage control through area division are as follows: Simplifies the voltage control problem, reduces the number of control actions, guides operators on decision making, allows obtaining stable voltage limits, minimizes exchange of reactive power between neighboring areas and enables centralized voltage control.

3 Proposed Methodology for Secondary Voltage Control Areas

The voltage stability analysis methodology by partitioning control areas is divided into two stages. The first one consists of the process of dividing the system and assigning pilot buses in each of the areas obtained. The second one consists in the analysis of the behavior of the system when voltage setpoints are set in the pilot buses according to different criteria for taking advantage of the nearby reactive power availability. The division into control areas is achieved by a hybrid method that considers two techniques widely studied in the literature. The first one is the partitioning technique based on the modal energy of the system. The second one is based on a MonteCarlo Simulation Approach, where the formulation of an optimization problem, that is supported by the concept of extended power flows, is performed to determine pilot buses, after which the areas related to each bus are found with K-means clustering.

3.1 Modal Energy Analysis for Control Areas Partitioning

This technique is based on the hypothesis that an extensive system can be fully analyzed and characterized from the observability and controllability of certain portions or areas of it [4] and [22]. The theory of this technique initially considers the reduced Jacobian that relates the voltage variations with respect to the reactive power variations in each bus, given by $\Delta Q = J_{QV} \cdot \Delta V$. In a modal space, given by the singular values of the Jacobian matrix, there are certain representative buses that provide almost total influence on the system, in addition, there are buses with a low influence. So, it is possible to divide a system into areas that can be controlled individually through the most representative bus or pilot bus [8]. The identification of control areas through modal energy is based on the theory of linear systems, in which the modes of this matrix, which are related to the most representative areas of the system, are identified. Then, the eigenvector matrix is determined to find participation factors and the assignment of the different buses to each area. According to [4], the modal energy of J_{QV} is given by

$$E(J_{QV}) = \sum\nolimits_{i=1}^{n} |\lambda_i(J_{QV})| = |\lambda_1|+|\lambda_2| + \ldots + |\lambda_p| + \ldots + |\lambda_n| \tag{1}$$

where λ_i is the $i-th$ singular value of the matrix J_{QV}. In the last expression, the singular values are ordered by magnitude, i.e., $|\lambda_1|> |\lambda_2| > |\lambda_p|> |\lambda_n|$. For the number of areas definition, the number of singular values which makes possible to capture a defined modal energy level is estimated. The following relationship is established.

$$E_{Np}(J_{QV}) = \sum\nolimits_{i=1}^{p} \frac{|\lambda_i(J_{QV})|^2}{||\Lambda||^2} = \frac{|\lambda_1|^2 + \ldots + |\lambda_p|^2}{|\lambda_1|^2 + \ldots + |\lambda_p|^2 + \ldots + |\lambda_n|^2} \tag{2}$$

where:

$$\Lambda = \begin{bmatrix} \lambda_1(J_{QV}) \\ \vdots \\ \lambda_n(J_{QV}) \end{bmatrix} \tag{3}$$

The index p is the number that corresponds to the control areas that allow meeting the condition of $E_{Np} > (\%)E_{min}$. This means that so many p areas must be chosen so that the energy captured by the singular values of greater magnitude exceeds a minimum energy level. Minimum power values of 75% to 99% are recommended to obtain optimal partitions of a network [4] and [23]. Next, the formulation for determining the control areas related with each pilot bus is outlined. Let the left eigenvector matrix be defined as follows:

$$\Psi_{ij} = \begin{bmatrix} \psi_{11} & \dots & \psi_{1n} \\ \vdots & \ddots & \vdots \\ \psi_{n1} & \dots & \psi_{nn} \end{bmatrix} \tag{4}$$

The eigenvector matrix is normalized as follows:

$$\eta_{ij} = \begin{bmatrix} \eta_{11} & \dots & \eta_{1n} \\ \vdots & \ddots & \vdots \\ \eta_{n1} & \dots & \eta_{nn} \end{bmatrix} = \frac{\Psi_{ij}^2}{||\Psi_i||^2} \tag{5}$$

where η_{ij} represents the relative participation of the $j-th$ bus in the $i-th$ mode. After this, the following participation matrix is defined, which results from the inner product between the normalized right eigenvector matrix η_{ij} and the percentage modal energy of each mode of the system. This is expressed by the following equations. Let the participation matrix of $j-th$ bus in total modal energy be defined as:

$$\eta_{Rij} = \frac{\left|\lambda_i\left(J_{QV}\right)\right|^2}{||\Lambda||^2} \times \eta_{ij} \tag{6}$$

The way in which each bus is assigned to the different areas represented by the singular values found is as follows. The values of the $j-th$ column of η_{Rij} are analyzed and the $i-th$ position of the greatest magnitude is identified. It means that the $j-th$ bus is assigned to $i-th$.area. With the steps outlined above, it remains to find the pilot bus for each area. The mathematical formulation is based on the criteria given in [4], which indicates that the pilot bus must have controllability and observability characteristics, that is, that they have the greatest influence on each of the buses. of the area and the least interaction with neighboring areas. Considering these criteria, the pilot bus in the $i-th$ area is the bus with the highest participation value in the matrix η_{Rij}.

3.2 MonteCarlo Simulation Approach

In the application of the Modal Energy Analysis, drawbacks can be found in certain operating states of the system since the reduced Jacobian may not faithfully represent the robust behavior of the pilot buses. For this reason, a second search method for pilot buses is proposed that considers non-analytical results, but simulation ones. The basic approach is to consider disturbances in a system for different operating states and find the minimization of an objective function (OF) that implicitly indicates which buses are the most robust. The formulation of the optimization problem is as follows:

$$\min_i OF = \min_i \sum\nolimits_{k=1}^{MC_{sim}} \frac{1}{n} \sum\nolimits_{i=1}^{n} \left(V_i^{base} - V_{i,k}\right)^2 \tag{7}$$

where n is the number of buses, V_i^{base} is the voltage in the $i-th$ bus corresponding to the base case and $V_{i,k}$ is the voltage in the $i-th$ bus after the perturbation considered in the k simulation of a total of MC_{sim} MonteCarlo simulations. For each simulation, the perturbation simulated is the stochasticity of every load according to a normal distribution function. The mean parameter is the actual Active Power (power factor is maintained constant). The standard deviation parameter is the 10% of the mean value.

The constraints defined in the optimization problem are determined so that the value of i considers only load buses. To jointly analyze the previous formulation with the concept of control areas, a previous partition of the system is developed with the use of grouping based on the K-means technique. To jointly analyze the previous formulation with the concept of control areas, a previous partition of the system is developed with the use of grouping based on the K-means technique. In this methodology, the distance matrix considered in the K-means algorithm is the reduced Jacobian of the system, where the elements of the main diagonal are replaced by zero. Because the reduced Jacobian represents the sensitivity between voltage and reactive power disturbances, it is appropriate to consider the concept that it is an electrical distance matrix.

In simulations, pilot bus options are tested for each area. The simulations consider an EPF and an optimal power flow, which maximizes reactive power reserves [24, 25]. EPF is a variation of the conventional Newton Raphson. In this case, the equations are modified to introduce the sensitivity of the voltage changes in pilot buses according to the control of remote elements, generally generators (view Fig. 2).

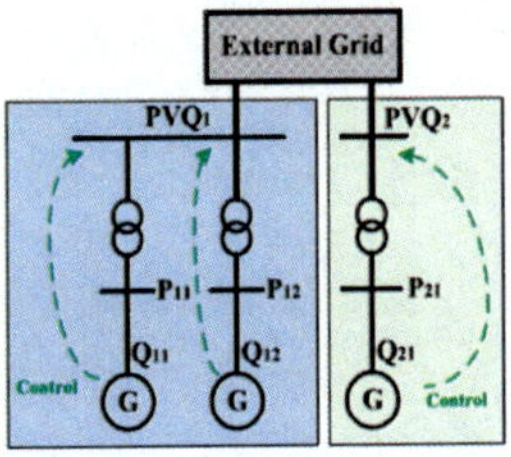

Fig. 2. Remote control of voltage in pilot buses

Although there is no analytical formulation of the optimization problem, the application of a heuristic algorithm to solve the problem is not necessary. The proposal is to collect the results of a significant set of Monte Carlo simulations to evaluate the most robust buses with these data. It can be understood that the application of this approach is more robust and does not lead to the application of an analytical formulation to solve the problem. The proposal is performed by means of simulations in PowerFactory with the development of an interface in engine mode with a Python code [26].

3.3 Control Areas Partitioning Validation

It is essential that in a real-time application, the partition of control areas with the modal energy technique is validated with some secondary technique that guarantees that the solution found is consistent. For this purpose, the methodology of the Residue for the transfer function between deviations in Reactive Power and deviations in the corresponding voltages. The validation is considered to verify that the sensitivities between voltage and reactive power enable a specific load bus to be considered as a pilot bus. Note that the remainder is analyzed as a transfer function given by:

$$G(s) = \frac{\Delta V_{Bus}}{\Delta Q} \tag{8}$$

The main idea of validation is to verify that the sensitivities between voltage and reactive power in the identified pilot buses are the highest for each area, that is, if the result corresponds to what was found with the methodologies used.

3.4 Secondary Voltage Control Proposal

Secondary voltage control requires setting a specific voltage in the pilot bus of an area in accordance with the injection of reactive power from the closest sources in the area. The purpose of this setting is that the voltages in the rest of the area are operating within safe margins. The available reactive power sources can be conventional generation, Wind Turbine Generators (WTG), Static Var Compensator (SVC), etc. Once the control areas have been defined, it is necessary to select which is the most critical bus in each area to improve its voltage profile. To determine the most critical bus in an area, the L-index, proposed in [16] is considered. The L-index stability index is a nodal type of metric whose calculation is easily obtained from the structure of the system (analysis of the Y_{bus} matrix), thus its calculation is fast. It provides the stability margin of voltage in steady state. This index for the $j - th$ bus is defined by:

$$L_j = \left| 1 - \frac{\sum_{i=1}^{N_G} F_{LG,ji} V_i}{V_j} \right| \tag{9}$$

where N_G is the number of generators in the area and F_{LG} is the matrix given by

$$F_{LG} = -(Y_{LL})^{-1} Y_{LG} \tag{10}$$

$$\begin{bmatrix} I_L \\ I_G \end{bmatrix} = \begin{bmatrix} Y_{LL} & Y_{LG} \\ Y_{GL} & Y_{GG} \end{bmatrix} \begin{bmatrix} V_L \\ V_G \end{bmatrix} \tag{11}$$

whereas L and G subscripts refer to load and generation buses respectively. On the other hand, in Secondary Control it is desirable that the reactive power contribution is efficient, that is, that WTG or SVC can be used if possible. With the support of Station Controllers in PowerFactory, secondary control is developed by comparing the optimization of reactive power reserves with arbitrary distribution instead.

4 Simulation Results and Analysis

The test system considered in this work is the IEEE 39 Bus-System model. The original model of this system is considered as a base case for the study. First, Control Areas might be identified. After that, the analysis for secondary voltage control is performed. Note that this network is modified in this work for considering High Wind Penetration.

4.1 Control Areas Partitioning

The Modal Energy Analysis and MonteCarlo Simulation Approach were applied in the IEEE 39 Bus-System for Identifying Control Areas. The control areas obtained with Modal Energy Analysis are shown in Fig. 3. The detail of each area is outlined in Table 1. In the application of the MonteCarlo Simulation Approach, a total of 5000 simulations were performed. The control areas obtained in this case are shown in Fig. 4. The detail of the control areas is shown in Table 2. According to the last results, a different number of control areas were determined with the methodologies applied. However, elements in Area 6 could be reassigned in another area due to their few elements.

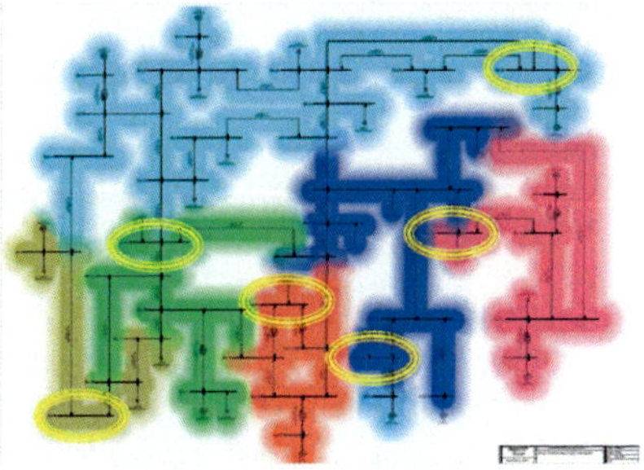

Fig. 3. Sketch map of control areas partitioning with modal energy analysis

Table 1. Detailed results of control areas partitioning with modal energy analysis

Area	Pilot Bus	Set of Load Buses	Set of Generation Buses
1	Bus 29	Bus 01, Bus 02, Bus 03, Bus 17, Bus 18, Bus 25, Bus 26, Bus 27, Bus 28, Bus 29	Bus 30, Bus 34, Bus 37, Bus 38
2	Bus 20	Bus 14, Bus 15, Bus 16, Bus 19, Bus 20, Bus 24	Bus 33
3	Bus 21	Bus 21, Bus 22, Bus 23	Bus 35, Bus 36
4	Bus 04	Bus 04, Bus 05, Bus 06	Bus 31
5	Bus 12	Bus 10, Bus 11, Bus 12, Bus 13	Bus 32
6	Bus 09	Bus 09	Bus 39

Since Modal Energy Analysis does not provide results as adequate for the operating point analyzed, in this case the results obtained with the MonteCarlo Simulation Approach are selected as priority. In this connection, the validation with the Residue Method is performed. In Fig. 5 the result of the validation is shown. The pilot buses that have been selected in each control area present the highest relationships given by the transfer function of the residue. In this way, the result of control areas obtained is considered as robust information for the secondary voltage control.

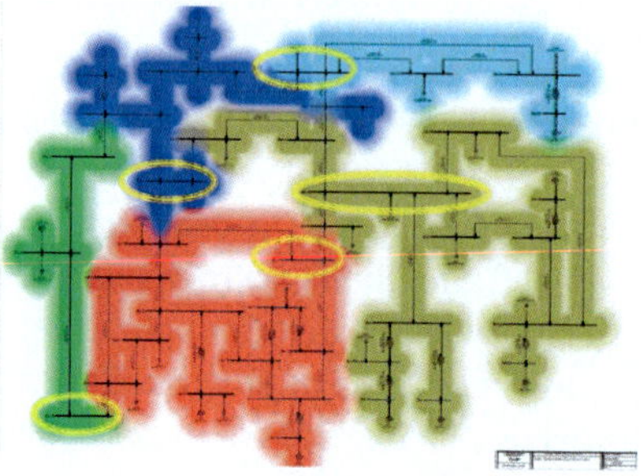

Fig. 4. Sketch map of control areas partitioning with the MonteCarlo simulation approach

Table 2. Detailed results of control areas with the MonteCarlo simulation approach

Area	Pilot Bus	Set of Load Buses	Set of Generation Buses
1	Bus 26	Bus 26, Bus 28, Bus 29	Bus 38
2	Bus 03	Bus 02, Bus 03, Bus 25, Bus 27	Bus 30, Bus 37
3	Bus 09	Bus 01, Bus 09	Bus 39
4	Bus 14	Bus 04, Bus 05, Bus 06, Bus 07, Bus 08, Bus 10, Bus 11, Bus 12, Bus 13, Bus 14	Bus 31, Bus 32
5	Bus 16	Bus 15, Bus 16, Bus 17, Bus 18, Bus 19, Bus 20, Bus 21, Bus 22, Bus 23, Bus 24	Bus 33, Bus 34, Bus 35, Bus 36

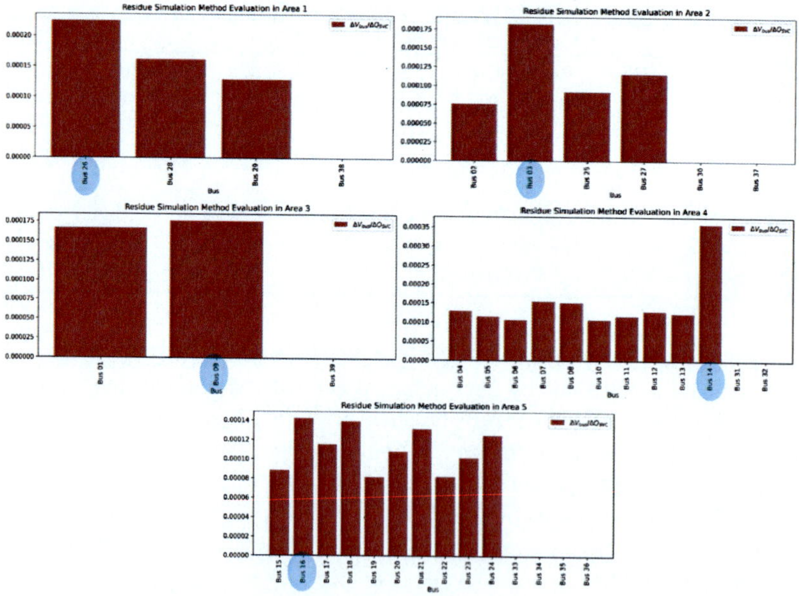

Fig. 5. Residue method validation for pilot bus selection with MonteCarlo approach

4.2 Secondary Voltage Control Analysis

The critical buses in each area were determined to identify possible stable operating margin violations. In Table 3 the critical buses are shown. Note that for Area 3, the pilot bus and the critical bus is the same bus. To consider a high penetration of wind energy in the system and analyze its possible contribution of reactive power in the secondary voltage control, the analyzed network is modified including a wind farm in each area. A high penetration of 250 MW is considered for each area. The WTG model considered is the Full Rated Converter (template of PowerFactory). This model allows to inject reactive power during fault conditions, in addition to providing a power ramp characteristic against transients. Moreover, to have a reactive power source that allows correcting adequately problems of voltage stability, a second modification in the analyzed system is considered. SVC devices are included in every pilot bus. Area 3 is not considered since it is a very small area. The resulting network is shown in Fig. 6.

Table 3. Representative buses identified for control voltage areas

Area	Pilot Bus	Critical Bus
1	Bus 26	Bus 29
2	Bus 03	Bus 25
3	Bus 09	Bus 09
4	Bus 14	Bus 07
5	Bus 16	Bus 21

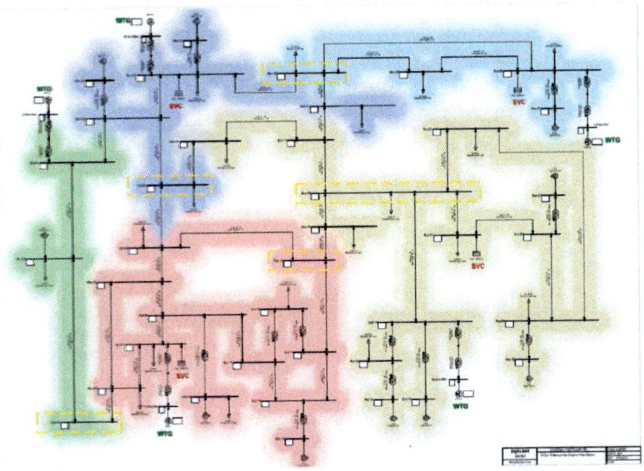

Fig. 6. Modified network considering the inclusion of WTG and SVC elements

To verify that SVC devices contribute positively to maintaining the voltages adequately controlled in the locations where these devices have been installed, Time-Domain simulations are performed to analyze the behavior of the dynamic voltage control. In this connection, a short-circuit event in Line 16–19 is simulated at t = 0.5 s during 100 ms. The results show an adequate behavior of the control (view Fig. 7). With the aim of modelling the stochasticity of wind power, Quasi-Dynamic simulations are performed considering variability of Wind Power as shown in Fig. 8. In addition, it must be considered that in actual systems voltage setpoints are not static. So, to consider different cases, quasi-static variability is introduced in the voltage references of the pilot buses.

Two simulation perspectives are considered for a period of 1 month of analysis (720 h). The first one is the network response without control and the second one is the inclusion of secondary control optimizing the reactive reserve in SVC, WTG, and conventional generators. Next simulations show conclusions about these perspectives.

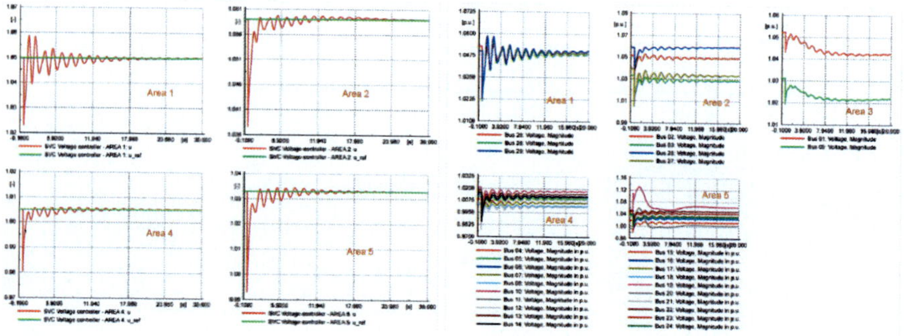

a) b)

Fig. 7. Dynamic voltage response: a) SVC elements and b) Bus voltages in areas

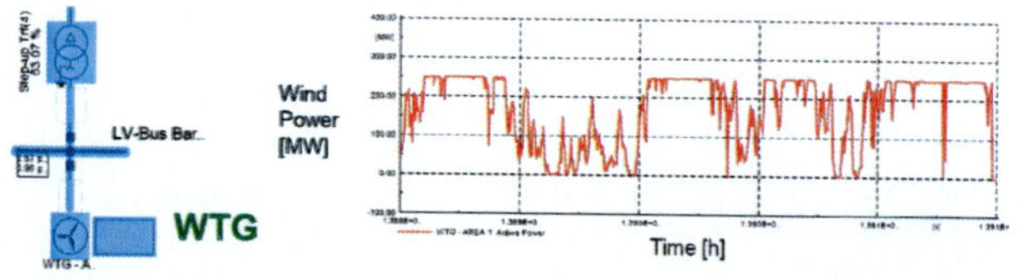

Fig. 8. Quasi-dynamic model of the Stochasticity in the WTG output power

An example of Quasi-Dynamic simulations in Area 4 is shown in Fig. 9. The analysis when there is no secondary control in reactive power sources and conventional generators yield power equally is included. On the other hand, simulations with secondary control in all sources are included for comparison purposes.

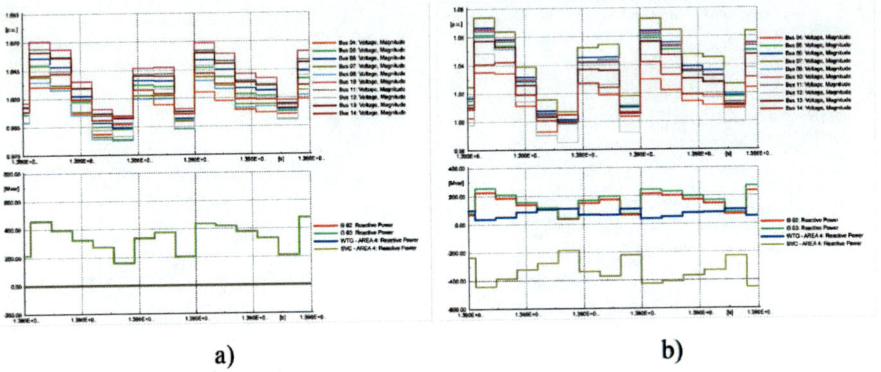

Fig. 9. Quasi-dynamic response for reactive power injection in area 4: a) Without contribution of WTG and SVC, b) With contribution of WTG and SVC

A statistical summary of the simulations is presented below. First, Fig. 10 shows the comparative result of the reactive power injections of the generators of the system. There is a better performance when considering secondary control in the operation of the network. For G 02 and G 03, the average power is reduced by half, for G02, G03 and G 08, the standard deviation decreases considerably. Next, the impact in bus voltages for the present analysis is shown in Fig. 11. A smaller standard dev. implies that the voltages remain more concentrated around the mean, so that it stays further away from an unstable condition. In this result it is observed that the optimal secondary control improves the voltage stability of the system, due to a less affected voltage profile.

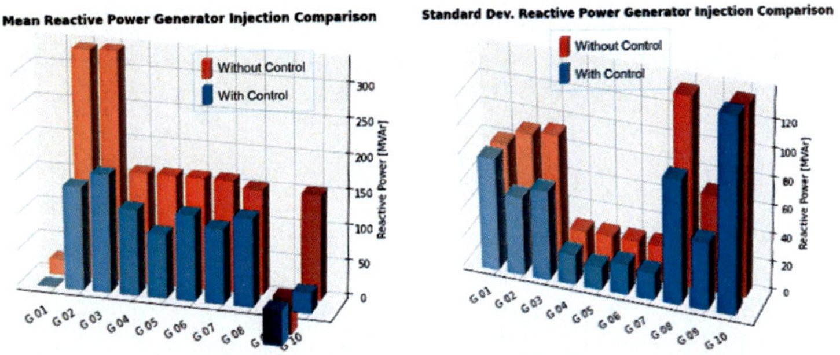

Fig. 10. Reactive power generator injection comparison with two perspectives

Finally, several seriously affected buses were identified when no effective secondary control is simulated. This is the case of the voltages obtained for Bus 07, Bus 19 and Bus 31, whose histograms are shown in Fig. 12. The lack of secondary control generates voltages outside of regions of safe operation more easily. In this result, when there is no control, the voltages exceed the allowable limits for transmission systems.

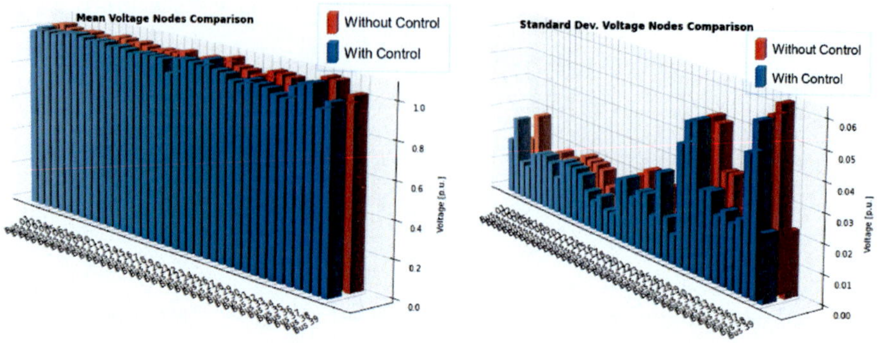

Fig. 11. Voltage buses comparison with two perspectives

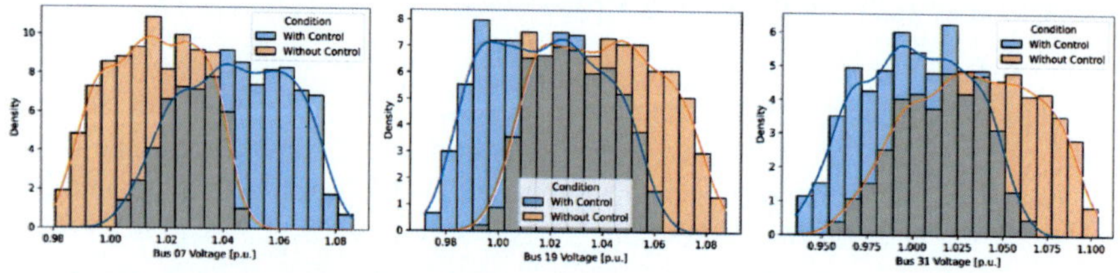

Fig. 12. Histograms of various buses severely affected when no control is considered

5 Conclusions

In this work a novel proposal for secondary voltage control based on hybrid methods was presented considering high penetration of Wind Power. The control was based on the identification of control areas and the determination of critical buses to incorporate reactive power compensation devices that contribute to secondary control. Several strategies to consider the concept of the extended power flow, the stochasticity of the wind power and the use of the reactive power reserve from various sources were proposed. The improvements that were evidenced in voltage control with the strategies proposed in this work are attractive to be applied in actual systems in which there are not commonly secondary control schemes that take advantage of reactive power resources and are not oriented to monitoring and define voltage setpoints only in the necessary buses. Note that the Modal Energy Analysis and the MonteCarlo Simulation Approach prove to be suitable for online applications. As shown in the analysis carried out, the definition of control areas will depend on the operating point. In this way, the application of hybrid methodologies for voltage control becomes even more necessary considering non-deterministic behavior of networks.

The control areas and pilot buses determined for the IEEE 39 Bus-System are somewhat similar to those determined in several research works. However, differences appear because this work considers the uncertainty of wind and a more proper detail of the generation response and stochasticity of the load. Furthermore, the considered hybrid methods allow the required degree of controllability in pilot buses. Compared with the methodologies in the state of the art, it is concluded that the proposal of this work can address the following needs that have not been considered in previous works:

- A methodology that considers simulations which provide more detail in the generator response including wind power and load levels.
- The inclusion of extensive and representative number of scenarios (operating points), which consider different load levels, uncertainty in the wind power and alternative reactive power sources such as SVC elements.
- A methodology that is available to be applied in a general way in actual systems.

Finally, to extend the analysis to a wider field of application, the inclusion of an optimal hierarchical voltage control, which considers more renewable energy sources, is currently being investigated.

References

1. Bedoya, A.: Metodología para el análisis de estabilidad de tensión mediante la división de redes en áreas de control. Universidad Nacional de Colombia, Medellín (2014)
2. Morales, C.: Estudio comparativo de metodologías para la detección de áreas de control de tensión. Universidad Tecnológica de Pereira, Pereira (2016)
3. Pengwei, D., Ross, B., Tuohy, A.: Integration of Large-Scale Renewable Energy into Bulk Power Systems From Planning to Operation. Springer, New York (2017)
4. Rios, J., Zamora, A.: Secondary Voltage Control Areas through Energy Levels. In: 2016 IEEE PES Transmission & Distribution Conference and Exposition - Latin America (PES T&D-LA) (2016)
5. Conejo, A., de la Fuente, I.: Goransson, S.: Comparison of alternative algorithms to select pilot buses for secondary voltage control in electric power networks. In: Proceedings of MELECON '94. Mediterranean Electrotechnical Conference, Antalya (1994)
6. Tangarife, C.: Estudio comparativo de metodologías para la detección de áreas de control de tensión. Universidad Tecnologica de Pereira, Pereira (2016)
7. Kessel, P., Glavitsch, H.: Estimating the voltage stability of a power system. In: IEEE Transactions on Power Delivery, vols. PWRD-1, no. 3 (1986)
8. Sanz, F., Ramirez, J., Correa, R.: Statistical estimation of power system vulnerability. IEEE (2013)
9. C.V. Voltage Stability and Controllability indices for Multimachine Power Systems: IEEE Trans. Power Sys. **10**(3), 1183–1186 (1995)
10. Fonseca, A., Pérez-Yauli, F., Salazar, G.: Loadability analysis based on short-circuit power. IET Gener. Transm. Distrib. **11**(10), 2540–2548 (2016)
11. La Gatta, P.O., Pereira, J.L.R.: An affine arithmetic method to identify voltage control areas. In: 2015 IEEE Eindhoven PowerTech. Eindhoven (2015)
12. González, X., Ramirez, J.: Control Jerárquico de Potencia Reactiva a paratir de una Formulación Multiobjetivo, CINVESTAV – Guadalajara – MEXICO, Guadalajara (2018)
13. Yidan, L.: A Wide Area Hierarchical Voltage Control for Systems with High Wind Penetration and an HVDC Overlay. University of Tennessee, Knoxville (2017)
14. Conejo, A., Gómez, T., de la Fuente, I.: Pilot-bus selection for secondary voltage control. ETEP **3**(5), 359–365 (1993)

15. Soroudi, A., Amraee, T.: Probabilistic determination of pilot points for zonal voltage control. IET Gener. Transm. Distrib. **6**(1), 1–10 (2011)
16. Yan, W., Cui, W.: Pilot-bus-centered automatic voltage control with high penetration level of wind. IEEE (2015)
17. Popovic, D., Levi, V.: Extension of the load flow model with remote voltage control by generator. Electric Power Sys. Res. **25**, 207–212 (2012)
18. Qin, N.: Voltage Control in the Future Power Transmission. Springer, Aalborg (2018)
19. Kundur, P., Paserba, J., Ajjarapu, V., Canizares, C., Hatziargyriou, N.: Definition and classification of power system stability. IEEE Trans. Power Syst. **19**(2), 1388–1396 (2004)
20. Kundur, P.: Power System Stability and Control. McGraw Hill (1994)
21. Liu, C., Qin, N., Sun, K.: Remote voltage control using the holomorphic embedding load flow method. IEEE Trans. Smart Grid **10**(6), 1–15 (2019)
22. Amraee, T., Soroudi, A.: Probabilistic determination of pilot points for zonal voltage control. IET Gener. Transm. Distrib. **6**, 1–10 (2011)
23. Witten, I., Frank, E.: Data Mining: Practical Machine Learning Tools and Techniques. Elsevier, Oxford (2005)
24. Dragan, S., Levi, V.A.: Extension of the load flow model with remote voltage control by generator. Electric Power Sys. Res. **25**, 207–212 (1992)
25. Zhu, L., Zhou, S., Zhang, Y.: Extended Load-Flow Arithmetic for Voltage Stability Analysis. In: PowerCon 2000. 2000 International Conference on Power System Technology. Proceedings (Cat. No.00EX409). Perth (2000)
26. DIgSILENT GmbH: PowerFactory 2020 - Python Function Reference, Gomaringen: DIgSILENT GmbH (2020)

Study of Electromagnetic Fields Distribution in Tena Electrical Substation

W. P. Guamán(✉), N. P. Minchala, C. L. Velasco, X. A. Proaño, and G. N. Pesántez

Universidad Técnica de Cotopaxi, Latacunga, Ecuador
{wilian.guaman8956,nestor.minchala4704,carol.velasco7891, xavier.proano,gabriel.pesantes3889}@utc.edu.ec

Abstract. This article studies the distribution of electromagnetic fields (EMF) in medium and high voltage electrical substations. Using the method of reflected images, EMF is calculated analytically for incoming and outgoing lines, transformer bars in the substation, and aplying Matlab is visualized the distribution of EMF in a substation. Finally, the EMF distribution in the principal emission sources (transformer lines and busbars) is simulated using QuickField software. For this reason, a comparative analysis is made between the data obtained from the simulation and the calculations, to contrast these results with the information available in the EMF measurement database and the technical characteristics of the substations belonging to the National Transmission System.

Keywords: Electric field · Magnetic field · Electrical substations · Power transformer · Power systems

1 Introduction

Every electric charge generates an electric field and every moving electric charge generates a magnetic. When electricity is generated, transported, distributed, or used, electricity is produced fields and low-frequency magnetic fields (50 Hz or 60 Hz). Then, the electric field is related to the system voltage, this does not change significantly with time, its variation occurs only between a distance and the source, while the magnetic field varies both in time and space, being related to the currentfield [1].

The substation is one of the most important elements to the operation and reliability of the electrical power system and is used as a connection and switching point for transmission lines, distribution feeders, power generation circuits, and transformers. However, these elements produce electric and magnetic fields of industrial frequency, the magnitude of these fields will depend on several factors such as geometrical, design, working voltage, power, arrangement of the elements, number of conductors, the distance between conductors, among others [2].

M. Botto-Tobar et al. (Eds.): *Latest Advances in Electrical Engineering, and Electronics*, LNEE 933, pp. 33–46, 2022.
https://doi.org/10.1007/978-3-031-08942-8_3

The study developed in [3] presents the results of magnetic field measurements carried out in electrical substations with voltage levels between 115 kV and 230 kV, as well as those presented during the works carried out on the system energized in 13.2 kV circuits of Empresa de Energía del Pacífico S.A. in Colombia. A comparative analysis is performed between the values measured in the field and the limits established by local regulations.

In Ecuador, [4] conducted a study of the environmental impact produced by electrical substations to determine the levels of electric and magnetic fields to which workers are exposed inside and around the substation in 2012. Electric Field Intensity, Magnetic Field Intensity and Magnetic Flux Density were measured in accordance with the provisions of the Non-Ionizing Radiation Standard for Electromagnetic Fields, and points were selected located on the boundaries, vertices of the properties and near the emitting sources. The electric field measurement was taken at a height of 1.5 m from the ground and at a minimum distance of 2.5 m from the generating sources, while for the magnetic field, the measuring equipment was installed at a height of one meter and at a distance of 2.5 m from the generating source.

There are no recently published studies that provide evidence of the evolution of EMF due to the increase in demand in electrical substations in Ecuador. Therefore, this work constitutes a reference base case for future works, although the authors only analyzed the Tena substation. The EMF estimation will be analytical, applying the methodology of the UNE-CLC/TR 50453 for transformers and EPRI for transmission lines in a calculation tool developed in Matlab. Then, we use the Quickfield software to calculate the EMF with the finite element method, to finally contrast the calculated, simulated and measured results in the field. So, if the error taking as reference the measured data does not exceed 5% on average, the tool is validated and could be applied to estimate the EMF with new substations.

2 Materials and Methods

2.1 EMF in Substations

Substations are made up of important elements that require periodic inspection and maintenance. The operators performing these tasks are exposed to significant electric and magnetic fields; EMF's outside electrical substations are usually more intense. The sources that cause significant magnetic fields are busbars, incoming and outgoing lines, power transformers, and yard equipment [5].

Typical magnetic field emission values for transmission substations from 245 kV to 400 kV are 10 μT, while for distribution substations from 11 kV they are 1.6 μT [6].

Due to the different nature of the elements and either by their construction or by the function they perform, they are classified into 3 categories of EMF emitting elements, the first category is the overhead lines and circuits that connect to power transformers and busbars, in second place, there are the subway lines or conduits, finally, the power transformers appear. In this classification, measurement and protection cables are not considered EMF sources because the magnitude of the current flowing through them is in the order of 100 to 300 times lower than the currents flowing through power circuits [7].

2.2 EMF Calculations

For the calculation of electric fields, a Matlab interface was developed, based on the calculation methodology established by EPRI [8] for transmission lines and UNE-CLC/TR IN 50453 [9] for power transformers.

Incoming Lines, Outgoing Lines and Busbars

The electric field generated by power transmission lines can be calculated using a simplified two-dimensional analysis, applying the following considerations [8]:

- The charges are only on the surfaces of the conductors.
- The conductors of the lines can be simulated as a set of cylindrical conductors parallel to each other.
- The dielectric medium between the conductors and the ground is equal to the permeability of the free space.
- The charges distributed on the surface of a conductor are simulated by a charge placed in the center of the conductor.

The earth's surface charges were simulated with image charges of equal magnitude, but of opposite polarity to the conductor charges. These are placed below the earth's surface as if it were a perfect mirror reflecting the conductors [10]. Each phase is reduced to a single conductor whose equivalent diameter is calculated with the following equation:

$$\mathrm{d_{eq}} = \mathrm{d_b} \cdot \sqrt[n]{\frac{\mathrm{n} \cdot \mathrm{d}}{\mathrm{d_b}}}. \tag{1}$$

$$\mathrm{d_b} = \frac{S}{\mathrm{sen}\left(\frac{\pi}{n}\right)} \tag{2}$$

where, d_b is the diameter of the bundle of conductors, n is the number of sub conductors in the bundle, d is the bundle diameter and d_{eq} is the equivalent diameter.
Then, to establish a linear relationship between the potentials of the conductors and each of the charges associated with them, in a linear system where there is a set of n conductors in a dielectric medium, the potential coefficients P_jm are determined, which will depend on the arrangement of the conductors in the structure, having the following configurations:

Flat Horizontal Configuration

The potential coefficients for a system of parallel conductors (see Fig. 1) have simple expressions:

$$P_{AA} = \frac{1}{2\pi\varepsilon} \cdot \ln \frac{4H_A}{d_{eq}} \tag{3}$$

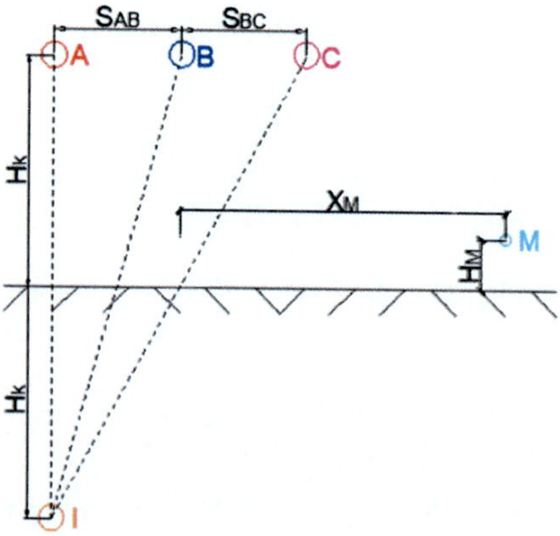

Fig. 1. Conductors and their images.

In addition, the mutual coefficients are calculated using the following equation:

$$P_{AB} = \frac{1}{2\pi\varepsilon} \cdot \ln \frac{s'_{AB}}{S_{AB}} \tag{4}$$

For this provision is met:

$$P_{AA} = P_{BB} = P_{CC} \tag{5}$$

$$P_{AB} = P_{BA} = P_{BC} = P_{CB} \tag{6}$$

$$P_{AC} = P_{CA} \tag{7}$$

In addition, there are two other typical configurations of transmission lines, in Fig. 2 there are the vertical and delta configurations respectively.

The coefficients form a square matrix called the array of the Maxwell potential coefficients (P).

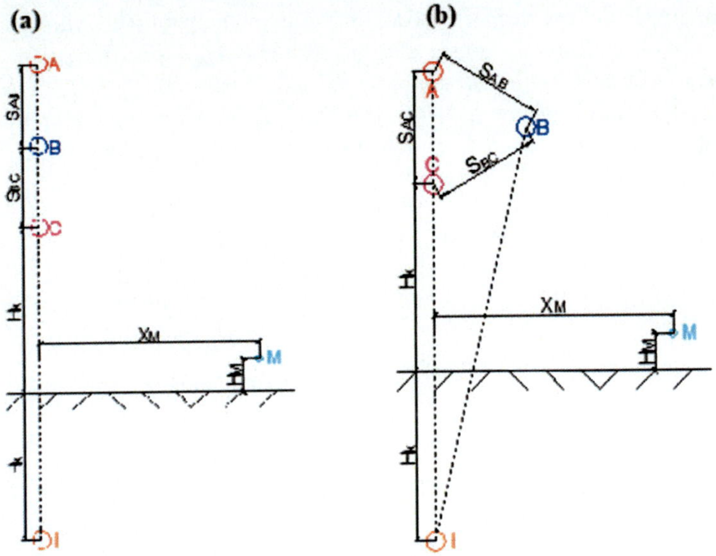

Fig. 2. Vertical and Delta configuration in transmission lines.

$$[P] = \begin{bmatrix} P_{AA} \; P_{AB} \; P_{AC} \\ P_{BA} \; P_{BB} \; P_{BC} \\ P_{CA} \; P_{CB} \; P_{CC} \end{bmatrix} \tag{8}$$

The electric field at a point in space can be calculated from the Q charges. The charges are calculated by knowing the voltage, V_k, applied to each conductor. Then:

$$[Q] = [P]^{-1} \cdot [V] \tag{9}$$

$$[C] = [P]^{-1} \tag{10}$$

The RMS value of the real and imaginary components, horizontal and vertical, E_{rx}, E_{ix}, E_{ry} and E_{iy}, fully characterize the vector field. Therefore, the RMS value of the electric field is given by:

$$E_{rms} = \sqrt{E_{rx} + E_{ix} + E_{ry} + E_{iy}} \tag{11}$$

The following considerations should be taken into account when calculating the magnetic field:

- The magnetic field varies approximately in inverse proportion with the square of the distance from the center of the conductor configuration.
- The magnetic field is proportional to the line current.
- Up to a few hundred meters from the line, the earth return currents have a negligible effect compared to the currents in the line conductors [8].

$$B_{xi} = \frac{\mu_0 \cdot I_k}{2\pi} \begin{bmatrix} \frac{-(h_M - h_k)}{(X_M - H_K)^2 + (h_M - h_k)^2} \\ -\frac{(h_M + h_k)}{(X_M - H_K)^2 + (h_M - h_k)^2} \end{bmatrix} \tag{12}$$

$$B_{yi} = \frac{\mu_0 \cdot I_k}{2\pi} \begin{bmatrix} \frac{X_M - H_K}{(X_M - H_K)^2 + (h_M - h_k)^2} \\ -\frac{X_M - H_K}{(X_M - H_K)^2 + (h_M + h_k)^2} \end{bmatrix} \tag{13}$$

The results magnetic field are:

$$B = \sqrt{B_{rx} + B_{ix} + B_{ry} + B_{iy}} \tag{14}$$

EMF in Power Transformers

The transformer generally consists of two adjacent windings of conducting wire wound around a single core of magnetic material. Therefore, transformers are machines that base their operation on the principle of magnetic induction, which states that a conductor through which circulates a time-varying electric current creates around it an equally variable magnetic field [11].

The report Evaluation of Electromagnetic Fields around Power Transformers [9]. gives guidance for the evaluation of electromagnetic fields around power transformers.

The methodology proposed in this document relates the three-phase symmetrical currents flowing through the bushings of the transformer with the magnetic field generated, so for the measuring point M the magnetic field has a value of:

$$B_x = \frac{(x+d) \cdot I_A}{(x+d)^2 + y^2} + \frac{y \cdot I_B}{x^2 + y^2} + \frac{(x-d) \cdot I_C}{(x-d)^2 + y^2} \tag{15}$$

where x, y and z are the value of the coordinates of the measuring point M, I the RMS current value per bushing, and d is the distance between the bushings (see Fig. 3), if the busbars assume a given length, the magnetic field is given by the following expression:

$$B_{\text{Tot}} = 2 \cdot 10^{-7} \cdot I \cdot \left(\frac{\sqrt{3} \cdot d}{1 + d^2} \right) \cdot \text{sen}(\propto) \tag{16}$$

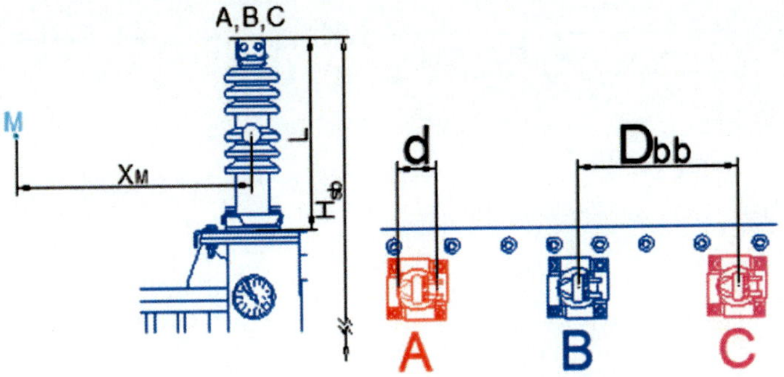

Fig. 3. Distances considered for the calculation of EMF emitted by the Transformer [9].

3 Analysis and Discussion of Results

This section details the results obtained in the calculation and simulation of the proposed elements, these data were validated with the available data from a previous measurement of substations of the national transmission system, managed by CELEC EP.

Using the interface developed in Matlab, the electrical parameters, arrangement of the elements, and the distance to reference the measuring point are entered manually. Then the program uses that information for the EMF calculation at different distances, obtaining, as a result, the graph of the magnitude of the electromagnetic fields emitted by the incoming and outgoing lines and busbars of the substation.

On the other hand, QuickField software will allow simulating and analyzing the distribution of electromagnetic fields in lines and transformers using the finite element method. The process starts by defining the problem and its parameters, to draw the geometrical model. Then, physical values must be assigned to the elements of the problem and the mesh must be created. Finally, the simulation can be run and the results obtained for further analysis.

3.1 EMF Simulation on Incoming and Outgoing Lines

EMF measurements obtained for the Tena substation, located in the province of Macas - Ecuador, with a power of 33 MVA and 138/69 kV, are taken as a reference. The configuration of the incoming and outgoing lines is delta type, single circuit. While the busbars are simple horizontal Table 1 and Table 2 shows the results obtained with the tool developed in Matlab and the percentage of error concerning the measured values.

In addition, the tool allows visualizing the behavior of the electric and magnetic field recorded at 1 m above ground level. The results obtained on the outgoing 69 kV line are shown below (Fig. 4 and 5):

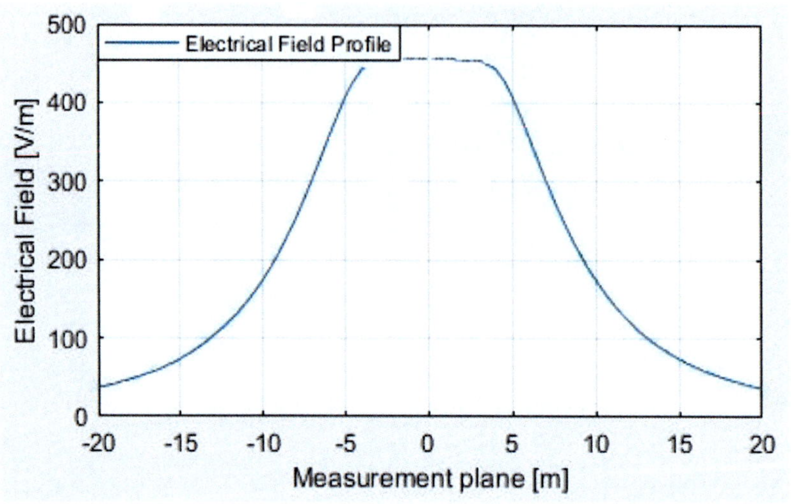

Fig. 4. Electric Field of the 69 kV busbar - Matlab interface.

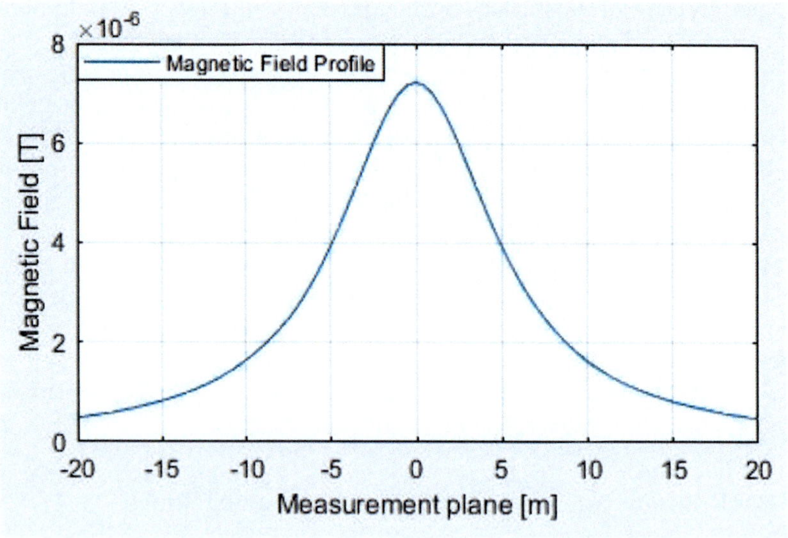

Fig. 5. Magnetic Field of the 69 kV busbar - Matlab interface.

Table 1. Electric field intensity data.

Line type	Measure [V/m]	Calculation [V/m]	Mistake %	Simulation [V/m]	Mistake %
Input substation	401,910	403,772	0,463	401,480	0,107
Busbar 138 kV	477,900	467,421	2,193	474,350	0,743
Busbar 69 kV	430,590	418,652	2,772	430,020	0,132
EEASA output	291,490	287,121	1,499	302,540	3,791

Table 2. Magnetic field intensity data.

Line type	Measure [uT]	Calculation [uT]	Mistake %	Simulation [uT]	Mistake %
Input substation	4,070	3,914	3,833	3,910	3,833
Busbar 138 kV	5,170	5,113	1,1103	4,900	5,222
Busbar 69 kV	6,520	6,267	3,880	6,340	2,761
EEASA output	4,980	4,900	1,586	4,970	0,201

It is observed that, at the measurement points closest to the lines, both the electric field and the magnetic field reach their highest value and as this point moves away from the center the EMF decreases. If the measurement point is replaced by 2.5 m for the 69 kV bus, the values obtained are 418.652 V/m and 6.267 uT for the electric and magnetic fields respectively.

3.2 Simulation of EMF in Substation Busbars

On the other hand, Fig. 6 and Fig. 7 shows the simulation of the electric and magnetic field of the 69 kV busbars using QuickField software to identify the EMF intensity present around the space surrounding the conductors. It is evident that, as in the incoming and outgoing lines, the electric and magnetic fields take their highest value when they are closer to the emitting sources, which are the conductors.

3.3 Simulation of EMF in Substation Busbars

There are two factors in addition to voltage and current that influence the electric and magnetic field, the first is because the transformer is designed in the most optimal way possible to reduce operating costs so that the amount of EMF that is scattered from the core is the smaller. The second is due to the shielding effect whereby the field across the metal transformer core is reduced [5].

The most important value of the magnetic field is the one generated in the low voltage bushings because the current flowing through these bushings is higher than the high side.

The influence of the electric field at the substation is of the same type as the influence of the electric field near the transmission line. When designing a new substation, the electric field of existing substations is a useful reference. The electric field at one meter above ground can be easily described by the electric field contour line drawn on the substation plan. The contour map is a convenient way to show the distribution of the ground electric field in the substation area.

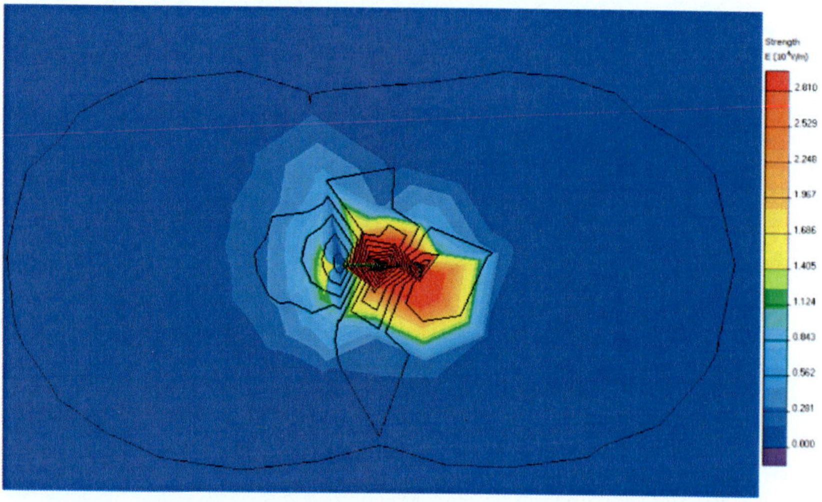

Fig. 6. Electric Field distribution of the 69 kV busbar - QuickField.

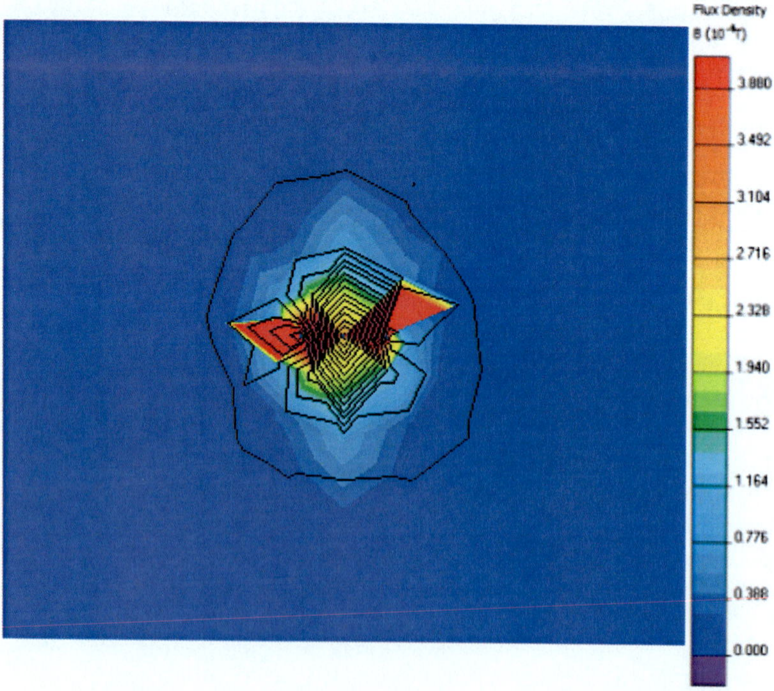

Fig. 7. Magnetic Field distribution of the 69 kV busbar - QuickField.

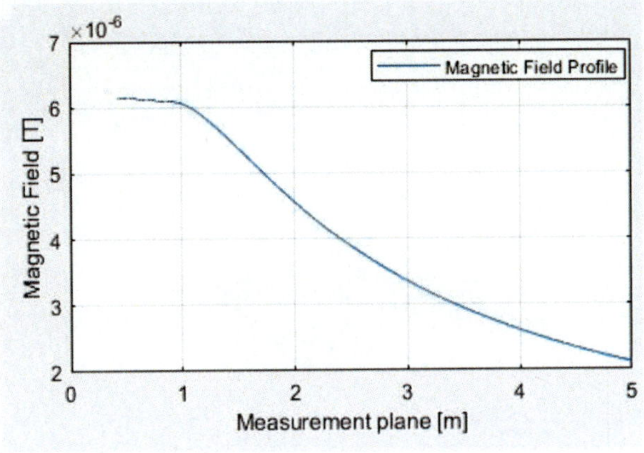

Fig. 8. Magnetic field for the 138 kV side of the transformer - Matlab Interface.

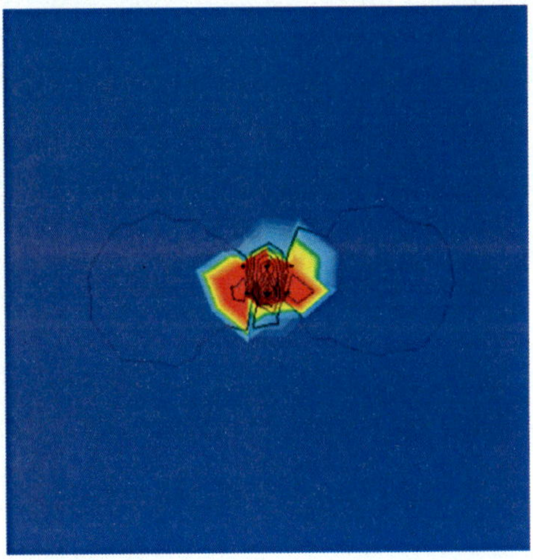

Fig. 9. Magnetic field in the transformer bushings - Quikfield.

Figure 8 and Fig. 9 show the electric field distribution throughout the substation. It can be seen that in the yard corresponding to the 138 kV system, the intensity is higher than in the 69 kV yard. However, it increases as it approaches the power transformer because this equipment has its energized parts closer to the ground.

Furthermore, the magnetic field distribution is shown in Fig. 10, where the magnitude increases in the 69 kV yard compared to the 138 kV yard, contrary to the electric field distribution.

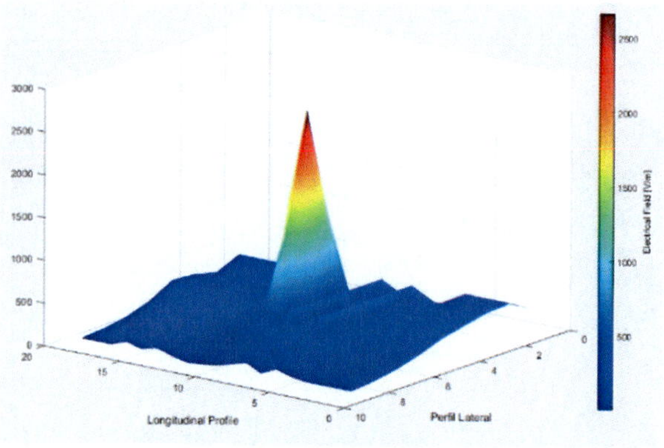

Fig. 10. Electric field for the 138 kV side of the transformer.

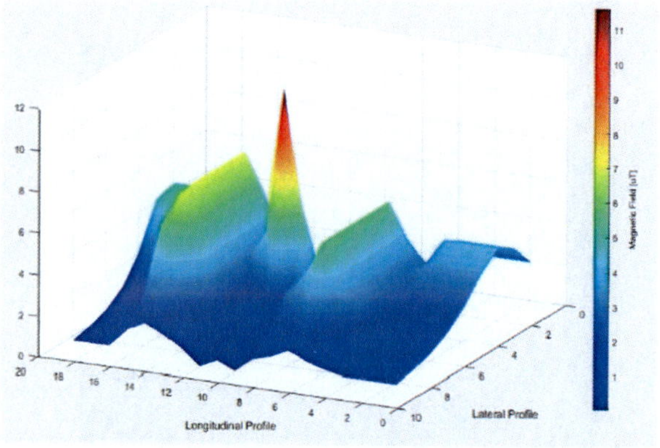

Fig. 11. Magnetic field in the transformer bushings.

Note that Fig. 10 and Fig. 11 only consider the four reference points indicated in Tables 1 and 2. Therefore, the results shown in both graphs for different measurement points do not necessarily correspond to the values that would be obtained at the Tena sub-station but are obtained by applying the least-squares approximation method.

4 Conclusions

The magnitudes of EMF measured, simulated, and calculated do not exceed those established by the Environmental Technical Regulations for the Prevention and Control of Environmental Contamination for the Infrastructure Sectors: Electric, Telecommunications and Transport, and by the International Commission on Non-Ionizing Radiation Protection international regulations, since the allowed values of electric field must be below 8333 V/m and for the magnetic field must be below 417 uT, presenting in the substations studied maximum values of 940.8 V/m in the electric field and 34.85 uT in the magnetic field.

For the calculation of the magnetic field, the amount of current flowing through the system at the time of the measurement is not known, so that, to contrast the results with the calculated values, it was necessary to vary the incoming current until was minimized the error (4.52%) between the calculated and measured values. Concluding that, for the incoming 138 kV line on the day of the study, a current of approximately 250 A was circulating.

Electromagnetic fields in a substation are distributed radially with high intensity at the EMF source and decrease as they move away. Of the elements analyzed in this work, it was found that the transformer is the EMF emitter because its energized parts are closer to the ground. In fact, for the Tena substation, the magnetic field produced by the transformer is approximately 39.153% higher than the highest value recorded in the 69 kV yard and 51.371% higher than the highest value recorded in the 138 kV yard. On the other hand, the electric field generated by the transformer is 84.947% and 78.452% higher than the field produced in the 69 kV and 138 kV systems, respectively.

References

1. Morales, J.A., Gavela, P.: Determinación de Campos Electromágneticos en Sistemas Eléctricos: Aplicación a casos reales, 1st edn. Abya-Yala, Quito (2015)
2. Cabezas, K., Jímenez, P., Ramirez, J., Canelos, R.: Simulación del campo electromagnético en una línea de transmisión de extra alto voltaje mediante el método de los elementos finitos. Revista Maskay **11**(1), 15–21 (2009)
3. Aponte, G., Escobar, H., Mora, A.: Evaluación del campo magnético al que están expuestos los trabajadores de subestaciones y circuitos energizados de las empresas de energía. Revista CIER **53**(5), 15–21 (2009)
4. Empresa Pública Estratégica Corporación Eléctrica del Ecuador - Unidad de Negocio TRANSELECTRIC: Estudio de Impacto Ambiental Definitivo Expost, EIAD Expost Subestaciones Eléctricas del Sistema Nacional de Transmisión que No Intersecan con el Sistema Nacional de Áreas Protegidas, Bosques y Vegetación Protectora y Patrimonio Forestal del Estado. CELEC EP, Quito, Ecuador (2012)
5. Leveque, R.: Método de cálculo y cumplimiento de normativa de campos magnéticos en subestaciones eléctricas. Trabajo de Fin de Máster. Universidad de Sevilla, Sevilla-España (2019)
6. Vielma, G.: Interacción De Campos Electromagnéticos de Extra Baja Frecuencia con el cuerpo humano, Mediciones de campo magnético en instalaciones de media tensión. Trabajo de Fin de Máster. Universidad de Chile, Santiago-Chile (2010)
7. Morales, J.A., Gavela, P., Bretas, A.: Electromagnetic fields in distribution feeders and electrical substations analysis: a study case in Ecuador. In: 2015 North American Power Symposium (NAPS), pp. 1–6 (2015)
8. Electric Power Research Institute: AC Transmission Line Reference Book-200 kV and Above, 3rd edn. EPRI, Palo Alto, CA (2005)
9. Asociación Española de Normalización y Certificación: Evaluación de los campos electromagnéticos alrededor de los transformadores de potencia. AENOR, Madrid, Spain (2008)

10. de Morais Sarmento, R.M.A.: Electric and magnetic fields in overhead power transmission lines. IEEE Latin Am. Trans. **10**(4), 1909–1915 (2012). https://doi.org/10.1109/TLA.2012.6272473
11. Gang, L., et al.: Simulation of the magnetic field distribution and voltage error characteristics of the three-phase three-component combined transformer with new three-cylinder core structure. IEEE Trans. Magn. **56**(4), 1–4 (2020). https://doi.org/10.1109/TMAG.2019.2945528

Mechanical Stress in Power Transformer Winding Conductors: A Support Vector Regression Approach

Fausto Valencia(✉), Hugo Arcos, and Franklin Quilumba

School of Electrical and Electronics Engineering, Escuela Politécnica Nacional, Ladrón de Guevara 253, Quito 170517, Ecuador
{fausto.valencia,hugo.arcos,franklin.quilumba}@epn.edu.ec

Abstract. A support vector regression model to determine the mechanical stress in winding conductors of a power transformer was developed in this research. Only the hyperparameter C, related to the regularization of the model, was varied during the design, thus simplifying the process and avoiding the need of using hyperparameter grids for performance comparison. The model was applied to a 400 MVA three-phase power transformer. The training and validation data were obtained from finite element simulations. The Python's library scikit-learn was used for the design process. The input data of the model were the currents circulating through the windings and the output data were the mechanical stresses of the conductors. Transient fault currents were used for the sample cases, with fault impedances that varied from 15 Ω to 80 Ω. Cases outside of the training data were used to test the model, giving accuracies from 1% to 20%. The greatest errors appeared for high fault impedances, i.e. for low fault current values. In a global analysis, it was seen that 99% of the errors were lower than 2.5%.

Keywords: Electromagnetic forces · Finite element method · Mechanical stress · Power transformers · Support vector regression

1 Introduction

The electromagnetic forces caused by the interaction between the electrical current circulating through the winding conductors and the magnetic field, could destroy the power transformer and end its life. For that reason it is recommended to perform a design review where the transformer capacity to withstand those forces must be clearly established [2,5].

Usually, these forces are determined using the finite element method (FEM) [4,13] to find the magnetic field in first place, and then the electromagnetic forces. For the magnetic field problem, FEM divides the medium into elements where the magnetic induction is considered constant [7]. In most of the problems, the elements are triangles and the vertices are nodes. Then, the

M. Botto-Tobar et al. (Eds.): *Latest Advances in Electrical Engineering, and Electronics*, LNEE 933, pp. 47–58, 2022.
https://doi.org/10.1007/978-3-031-08942-8_4

magnetic vector potential is determined at each node. The coordinates of these vertices constitute the discretization of the medium.

The problem with the FEM discretization, and with other less used methods with similar discretization such as the Finite Difference Method, is that each point represents a row and a column of a matrix. This matrix must be treated as part of an equation, i.e. it must be factorized and solved for each change in the condition of the problem [11]. The situation is even worse if it is considered that the best discretization must be found iteratively, mainly for nonlinear problems such as finding the magnetic induction in a ferromagnetic core [12].

As a consequence, FEM and electromagnetic force analysis have been done only for static cases. Dynamic studies, such as the fatigue analysis made by Araujo et al. [1], or the deformation analysis made by Zhang et al. [17], both in winding conductors, are particular studies that cannot be applied to real-time monitoring with the usual numerical methods.

Machine learning tools are an option to improve the speed in the solution of mechanical stresses. Some attempts have been made in regards to the magnetic field. For example, Mateev et al. use 2D images of the magnetic field to train a convolutional neural network [8]. This approach has been made for general geometries. However, in power transformers, there could be the problem of the dimensions in the medium, which vary from meters in the core to millimeters in the winding.

A highly accurate artificial neural network (ANN) model has been developed to directly determine the mechanical stress in winding conductors [15]. It uses the results of FEM simulations to train the network. The problem with ANN is the number of hyperparameters that must be tuned to have an acceptable model. A slight variation in one of the hyperparameter's value could strongly change the accuracy of the model [3]. For that reason, grids of hyperparameters are employed when designing ANN [16], which may confuse an inexperienced designer.

In this research, support vector regression (SVR) is used to design a machine learning model to find the mechanical stress in winding conductors. The design is based on the variation of only one hyperparameter, which strongly simplifies the process.

2 Method

The mechanical stress was calculated in a three-phase 400 MVA power transformer whose characteristics are shown in Table 1, and whose dimensions are indicated in Table 2. The transformer core material is US Steel Type 2-S 0.018 in. thickness.

The currents for the training data i_a, i_b and i_c are part of a transient fault according to (1) for the low voltage winding. ω is the angular frequency of the system, t is the time, ϕ is the angle representing the start of the fault, θ is the angle to Phase A (120° in a balanced system), $\lambda = r/l$, where r and l are the equivalent resistance and inductance seen by the fault. The fault impedance used

Table 1. Power Transformer under analysis

Variable	Value	Unit
Power	400	MVA
High voltage	230	kV
Low voltage	138	kV
Frequency	60	Hz
Group of connection	Yy0	
Impedance (own base)	33.61	%
Number of low voltage disks	100	u
Number of high voltage disks	105	u

Table 2. Transformer internal dimensions

Item	Length (m)
Core Diameter	0.9582
Core window height	3.034
Limb-limb separation	1.853
Low voltage winding inside diameter	1.108
Low voltage winding outside diameter	1.309
High voltage winding inside diameter	1.429
High voltage winding outside diameter	1.625
Low voltage disk height	0.0175
High voltage disk height	0.0136
Spacer block (located between disks)	0.010

to create the training data varies from $1\,\Omega$ to $5\,\Omega$ for the resistance and $15\,\Omega$ to $80\,\Omega$ for the inductive reactance.

$$\begin{aligned} i_a(t) &= I_{LV} \cdot [\sin(\omega t - \phi) + \sin(\phi) \cdot \exp(-\lambda t)] \\ i_b(t) &= I_{LV} \cdot [\sin(\omega t - \theta - \phi) + \sin(\theta + \phi) \cdot \exp(-\lambda t)] \\ i_c(t) &= I_{LV} \cdot [\sin(\omega t + \theta - \phi) + \sin(-\theta + \phi) \cdot \exp(-\lambda t)] \end{aligned} \quad (1)$$

2.1 Electromagnetic Forces and Mechanical Stress

The force per volume in a conductor with a current density **J**, immersed in a magnetic field **B** is given by (2) [6]. The magnetic field is related to the magnetic vector potential **A** by (3). In a homogeneous medium, where the permeability is μ, the magnetic Poisson's equation relates **A** to **J**, see (4). Therefore, to find the force acting on a conductor, the Poisson equation must be solved to find **A**, then **B** is calculated, and finally the force per unit volume is determined. For the solution of (4), the FEMM program was used [9].

$$\mathbf{f} = \mathbf{J} \times \mathbf{B} \tag{2}$$

$$\mathbf{B} = \nabla \times \mathbf{A} \tag{3}$$

$$\nabla^2 \mathbf{A} = -\mu \mathbf{J} \tag{4}$$

It was demonstrated that the radial force by itself causes the worst deterioration of the conductor [15] because the structure of the transformer nullifies the axial component effects of the force. Therefore, the mechanical stress was found considering only the radial component.

If a set of conductors that are part of the winding disk is considered as a ring with radius R_{ring}, the force P along the perimeter, acting on the cross section of the ring is given by (5). F_r is the radial force per length. The stress σ on the cross section S_c of the ring is calculated by (6) [14]. This procedure was implemented in Python, which interacts with FEMM to automate the algorithm [9].

$$P = F_r \cdot R_{ring} \tag{5}$$

$$\sigma = P/S_c \tag{6}$$

2.2 Support Vector Regression

The model takes the currents of the low voltage and high voltage windings as input data and delivers the mechanical stress in the winding conductors. Figure 1 shows a sketch of the internal architecture of the SVR model. The variables x_i represent the input data. They may become the support vectors sv_i according to its coordinate position in regards to a hyperplane. The determination of the stress is a nonlinear problem. Hence there are kernel functions k_i that handle the map transformation of the original relations. Finally, the resultant functions are added, and the stresses in the winding conductors are given.

The implementation of the SVR was made in the Python's library scikit-learn [10]. The kernel Radial Basis Function was chosen, and the hyperparameter C was modified to get the model behavior in regards to accuracy and regularization.

Before performing the training process, each pair of input/output data (x, y) was standardized to $(x_{\text{std}}, y_{\text{std}})$ by (7), according to the maximum and minimum values of the training input database $(x_{\text{max}}, x_{\text{min}})$, and of the training output database $(y_{\text{max}}, y_{\text{min}})$. Data standardization avoids that high values have much more influence than lower values in the designed model. For this research, only module standardization was performed. The data probability distribution stayed as original.

$$\begin{aligned} x_{\text{std}} &= \frac{x - x_{\text{min}}}{x_{\text{max}} - x_{\text{min}}} \\ y_{\text{std}} &= \frac{y - y_{\text{min}}}{y_{\text{max}} - y_{\text{min}}} \end{aligned} \tag{7}$$

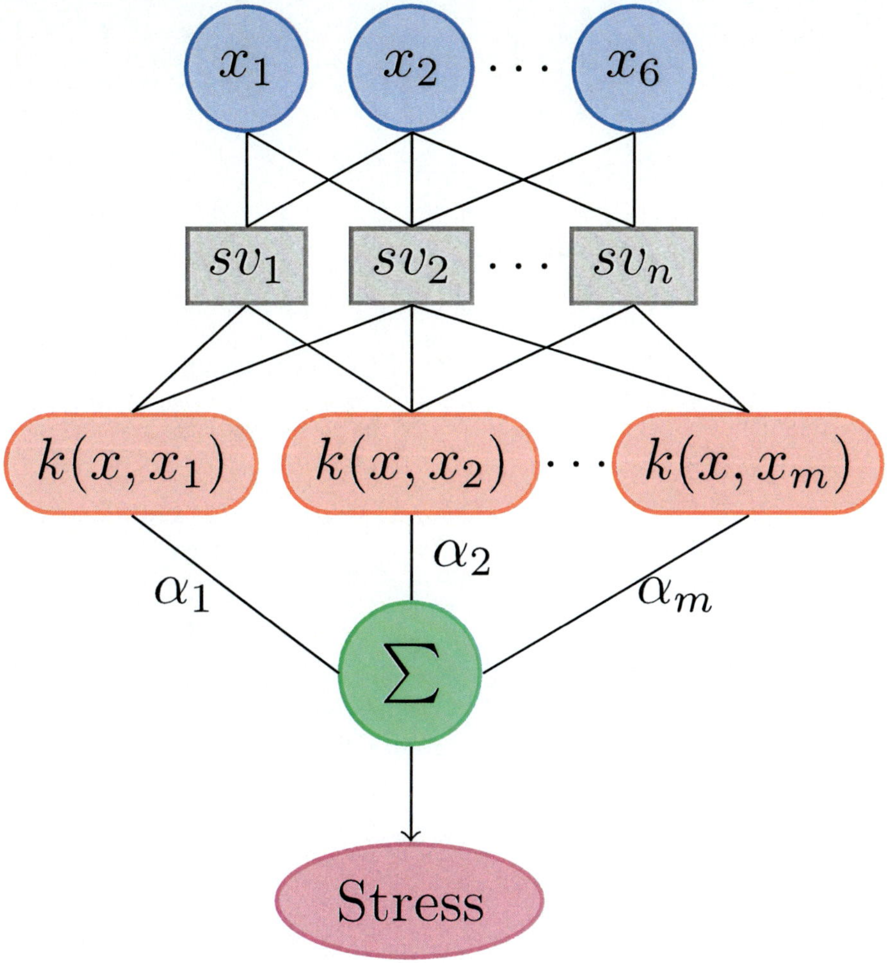

Fig. 1. SVR model architecture: x_i represents the input currents, sv_i is the support vector, and k is the kernel function.

The samples were divided into two groups to test against overfitting: 90% for training and 10% for validation data. Only the training data were used for the design and then the model was applied to the validation data. The best performance observed in the validation data was set as the final model.

The scikit-learn library determines the accuracy of the model with the variable R calculated by (8), where y_{true} is the true value of the output variable, y_{pred} is the output value predicted by the model, and $\bar{y}_{true}$ is the mean value of the output variable.

$$R^2 = 1 - \frac{u}{v}$$
$$u = \sum \left(y_{true} - y_{pred}\right)^2 \tag{8}$$
$$v = \sum \left(y_{true} - \bar{y}_{true}\right)^2$$

3 Results and Discussion

3.1 Results

As long as C was increased, the accuracy of the model was improved (Fig. 2). The lowest error was obtained when $C = 250$. After this point, the accuracy for the validation data began to increase, while the accuracy for the training data remained constant.

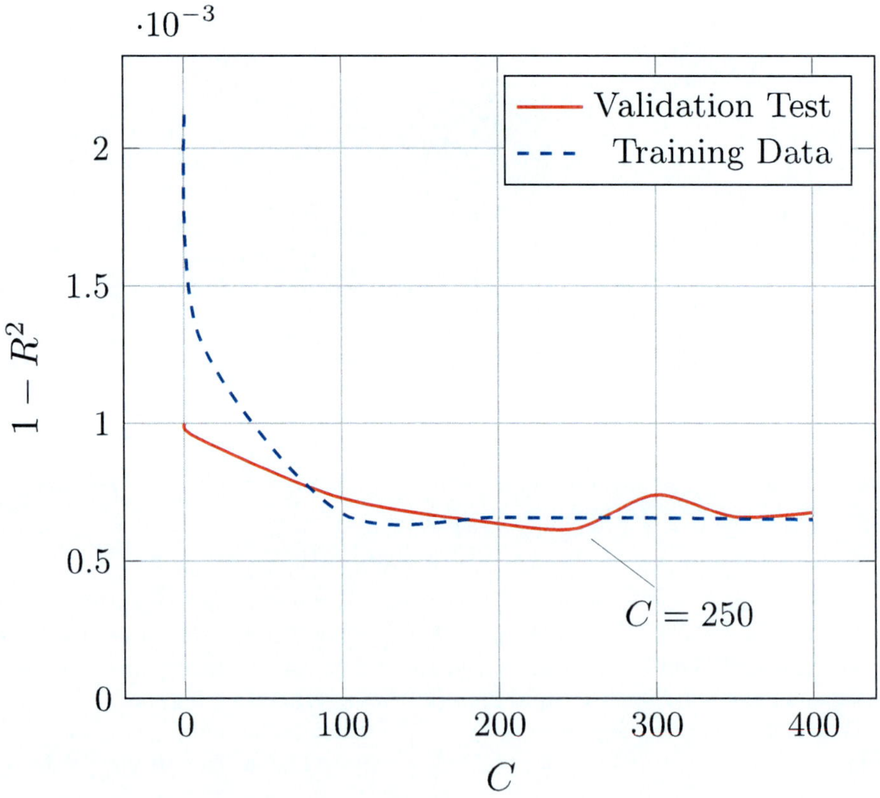

Fig. 2. SVR model accuracy according to the variations of C.

Figure 3 shows the Mean Absolute Percentage Error (MAPE) when the fault is located at the terminals of the power transformer, i.e. with a resistance of 1 Ω and an inductive reactance of 15 Ω. The errors had a value less than 1%. Phase B had the highest error and Phase A, the lowest one. The behavior of the low voltage and high voltage windings was similar.

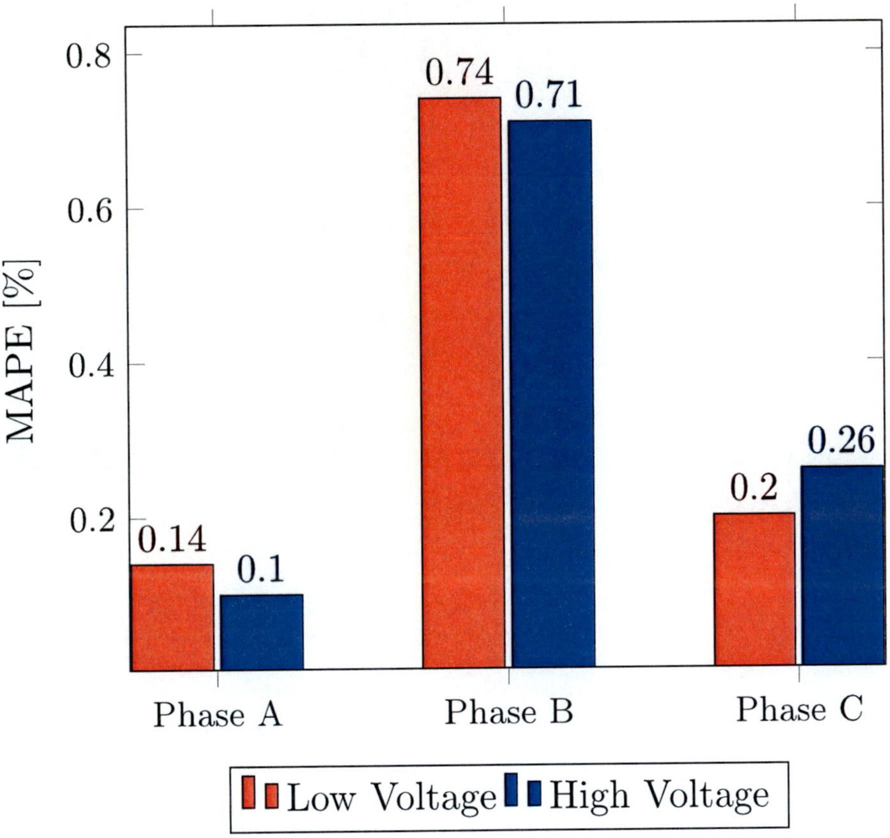

Fig. 3. Mean Absolute Error for a fault impedance $Z = 1 + j15\ \Omega$.

The MAPE for a fault with an impedance of 1 Ω of resistance, and 47 Ω of reactance is shown in Fig. 4. Phase B had the highest error with a value greater than 6%. Phases A and C had a similar behavior with errors around 2.8%. There is not too much difference between the results for the low and the high voltage windings.

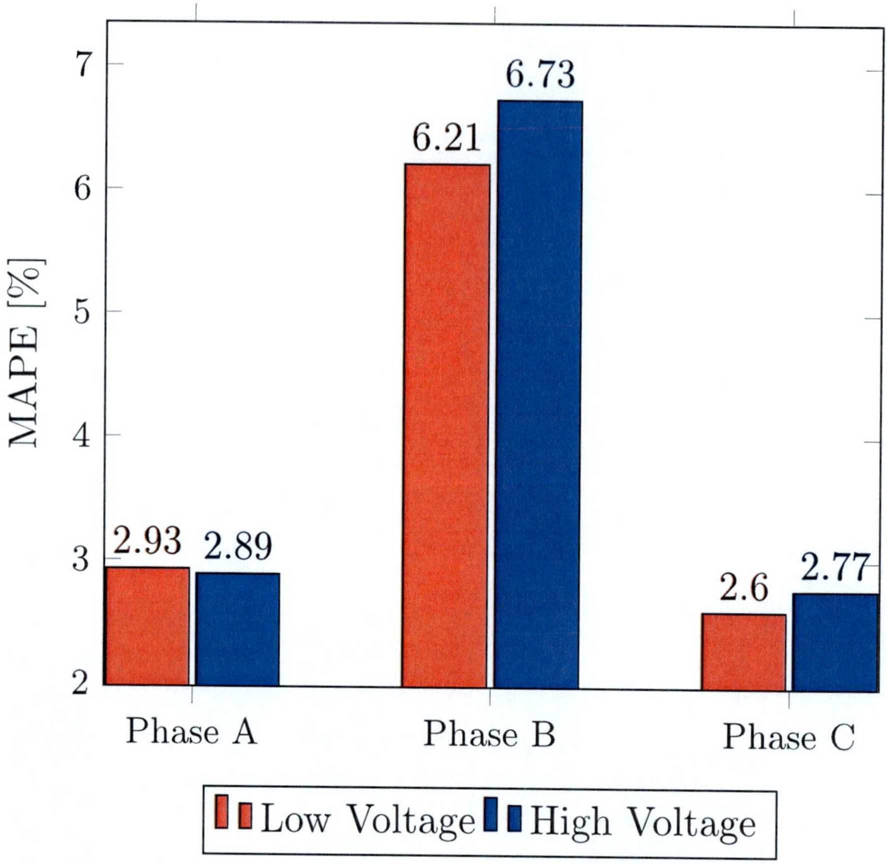

Fig. 4. Mean Absolute Error for a fault impedance $Z = 1 + j47\ \Omega$.

In Fig. 5 the MAPE for a fault with an impedance with a resistance of 1 Ω and a reactance of 80 Ω is shown. Phase B had the highest errors with values around 20%. Phases A and C had a similar behavior, rounding an error of 10%. The results for the high voltage winding for Phases B and C present a noticeable higher value than those of the low voltage winding. On the other hand, both windings have a similar error for Phase A.

Figure 6 shows the behavior of MAPE in relation to the value of the mechanical stress. The errors were less than 20% when the mechanical stress was greater than 10^5 Pa. The error increased as long as the mechanical stress was lower.

According to Fig. 7, 99% of the cases fell into the area of errors less than 2.5%. Moreover, 50% of the cases were located between 0.1% and 1.1%.

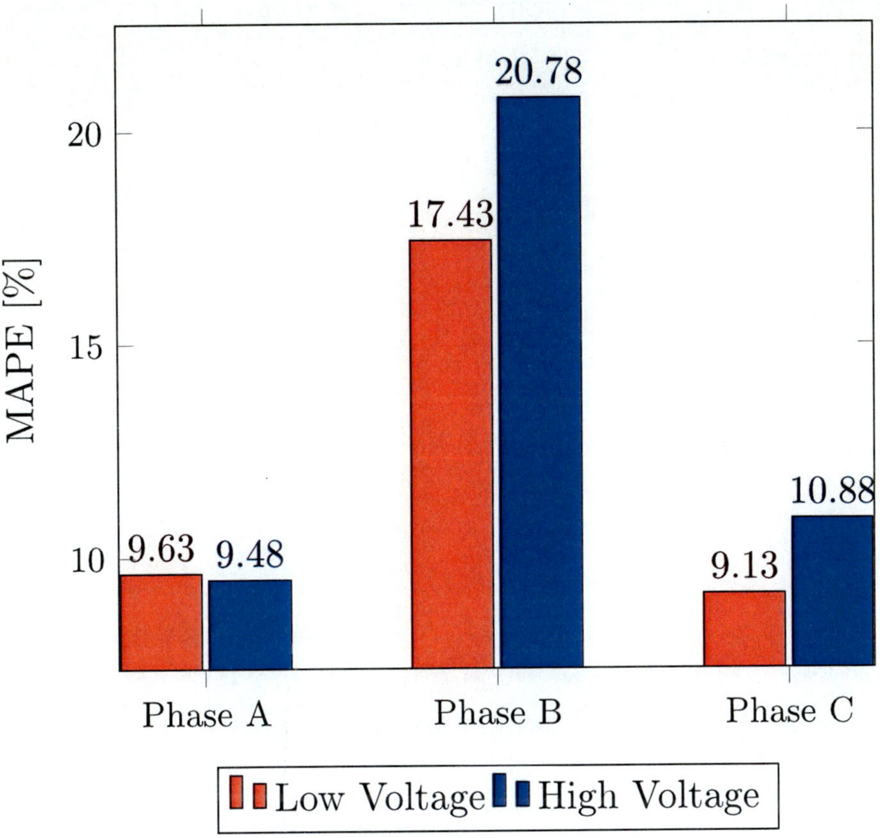

Fig. 5. Mean Absolute Error for a fault impedance $Z = 1 + j80\ \Omega$.

3.2 Discussion

The hyperparameter C shows the typical behavior of a model training. The validation test reduces the error until an optimal point, and then it begins to rise. On the other hand, the training data error stabilizes at a value and it is expected to be reduced if C is increased. Testing the model with both validation and training data is a common practice in machine learning, and avoids overfitting as it would have happened after $C = 250$.

In general, the accuracy of the model does not depend on the winding. Both low voltage and high voltage windings have similar errors.

The fault with the lowest impedance presents the best accuracy. This characteristic happens due to the high values of fault current and mechanical stress under those circumstances. The model is trained better for high values, even if they are standardized as it is the case for this research. This is confirmed in the relationship between the mechanical stress and MAPE.

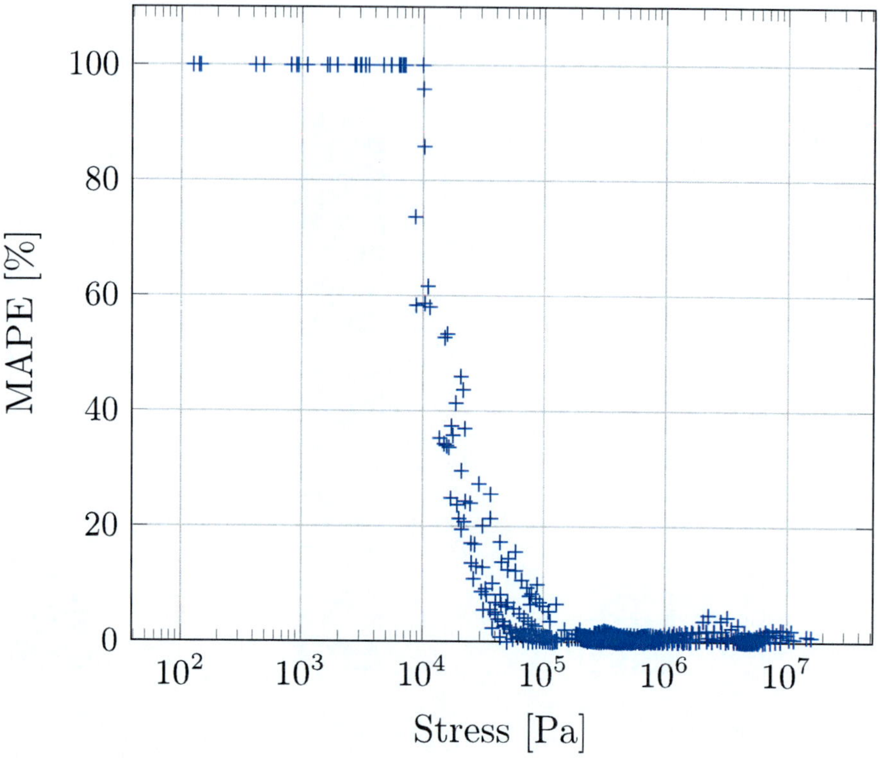

Fig. 6. Behavior of the MAPE in relation to the mechanical stress.

The accuracy presented for low mechanical stresses does not affect the global performance of the model. The high value errors appear for a small number of cases, as it is seen in the error distribution. These errors are the result of the transient period, where some peak currents are close to zero (as a minimum value). This means that the mechanical stress in those cases is lower than that of a normal operation condition of the power transformer. Furthermore, in a practical application, the user is more concerned about the effects of high mechanical stresses which can faster deteriorate the winding conductor.

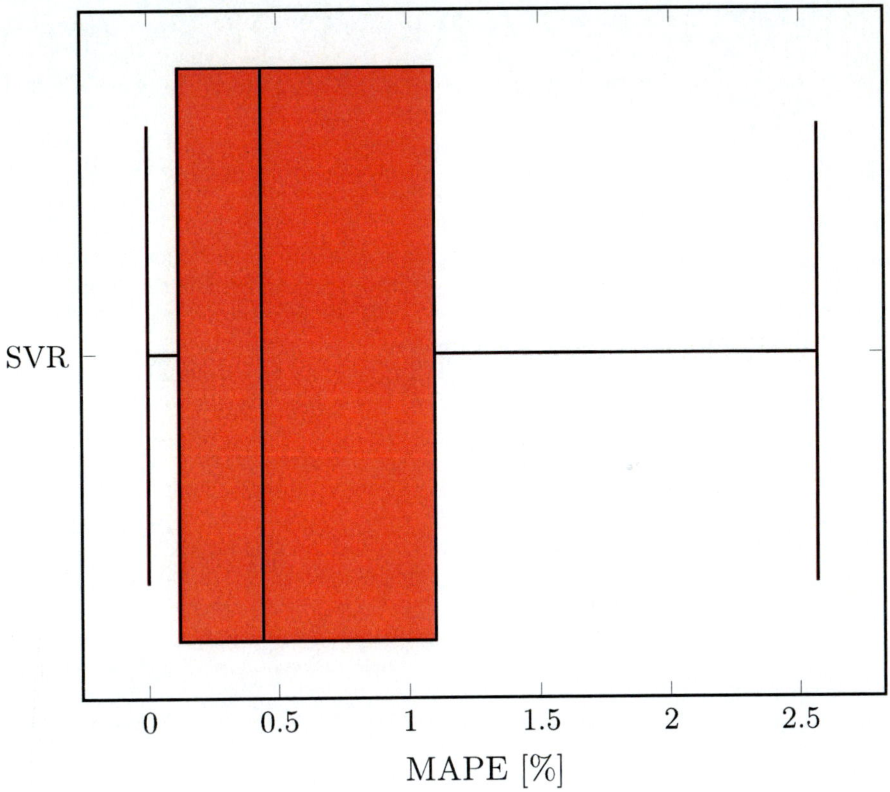

Fig. 7. Error distribution.

4 Conclusion

Support vector regression has proven to be accurate for the determination of mechanical stress in the winding conductors of a power transformer. Moreover, the design process is simple because only the regularization hyperparameter C needs to be varied. The highest errors are found for the lowest values of stress, which is not a huge problem because they represent less than 1% of the conditions with which the transformer must operate.

An additional analysis could be performed considering inrush instead of fault currents. The variation of permeability for this problem may constitute a new challenge when trying to obtain a high accuracy in the model.

References

1. Araujo, J.F., Costa, E.G., Andrade, F.L.M., Germano, A.D., Ferreira, T.V.: Methodology to evaluate the electromechanical effects of electromagnetic forces on conductive materials in transformer windings using the von mises and fatigue

criteria. IEEE Trans. Power Delivery **31**(5), 2206–2214 (2016). https://doi.org/10.1109/TPWRD.2016.2579165
2. Bertagnolli, G., Services, A.M.: The ABB Approach to Short-circuit Duty of Power Transformers. ABB, Zürich (2007)
3. Choi, D., Shallue, C.J., Nado, Z., Lee, J., Maddison, C.J., Dahl, G.E.: On empirical comparisons of optimizers for deep learning. arXiv preprint arXiv:1910.05446 (2019)
4. De Azevedo, A.C., Rezende, I., Delaiba, A.C., De Oliveira, J.C., Carvalho, B.C., De, S.B.H.: Investigation of transformer electromagnetic forces caused by external faults using fem. In: 2006 IEEE/PES Transmission & Distribution Conference and Exposition: Latin America, pp. 1–6. IEEE (2006). https://doi.org/10.1109/TDCLA.2006.311522
5. IEC: IEC 60075 Power transformers - Part 5: Ability to withstand short circuit. International Electrotechnical Commission, Geneva, Switzerland (2006). https://webstore.iec.ch/
6. Jackson, J.: Classical Electrodynamics. Wiley, Hoboken (2012)
7. Jin, J.M.: The Finite Element Method in Electromagnetics. John Wiley & Sons, Hoboken (2015)
8. Mateev, V., Marinova, I.: Machine learning in magnetic field calculations. In: 2019 19th International Symposium on Electromagnetic Fields in Mechatronics, Electrical and Electronic Engineering (ISEF), pp. 1–2 (2019). https://doi.org/10.1109/ISEF45929.2019.9096969
9. Meeker, D.: Finite element method magnetics Version 4.2: User's Manual. dmeeker@ieee.org, USA (2010)
10. Pedregosa, F., et al.: Scikit-learn: machine learning in python. J. Mach. Learn. Res. **12**, 2825–2830 (2011)
11. Rylander, T., Ingelström, P., Bondeson, A.: Computational Electromagnetics. TAM, Springer, New York (2013). https://doi.org/10.1007/978-1-4614-5351-2
12. Saitz, J.: Newton-Raphson method and fixed-point technique in finite element computation of magnetic field problems in media with hysteresis. IEEE Trans. Magn. **35**(3), 1398–1401 (1999). https://doi.org/10.1109/20.767225
13. Steurer, M., Frohlich, K.: The impact of inrush currents on the mechanical stress of high voltage power transformer coils. IEEE Trans. Power Delivery **17**(1), 155–160 (2002). https://doi.org/10.1109/61.974203
14. Timoshenko, S.: Strength of materials Part 1. Strength of Materials. D. Van Nostrand Company, Incorporated, New York (1940)
15. Valencia, F., Arcos, H., Quilumba, F.: Prediction of stress in power transformer winding conductors using artificial neural networks: Hyperparameter analysis. Energies **14**(14), 4242 (2021). https://doi.org/10.3390/en14144242
16. Young, S.R., Rose, D.C., Karnowski, T.P., Lim, S.H., Patton, R.M.: Optimizing deep learning hyper-parameters through an evolutionary algorithm. In: Proceedings of the Workshop on Machine Learning in High-Performance Computing Environments. MLHPC 2015. Association for Computing Machinery, New York (2015). https://doi.org/10.1145/2834892.2834896
17. Zhang, H., et al.: Dynamic deformation analysis of power transformer windings in short-circuit fault by fem. IEEE Trans. Appl. Supercond. **24**(3), 1–4 (2014). https://doi.org/10.1109/TASC.2013.2285335

Industrial Control and Automation

Development of an Industrial IoT Gateway Prototype

Diego Mancheno(✉) and Silvana Gamboa

Escuela Politécnica Nacional, Quito, Ecuador
{diego.mancheno,silvana.gamboa}@epn.edu.ec

Abstract. Nowadays, many industries have started to use different devices that allow the integration of process data through the internet to incorporate manufacturing into an Industry 4.0 scheme. In this regard, the present work involves the development of an IIoT Gateway to acquire data from industrial devices and through standard digital or analog inputs and outputs. The prototype provides bi-directional interaction with a remote station through Modbus RTU serial communications protocol under RS-485 interface. The implementation is based on an Arduino UNO board that incorporates an Ethernet module. The necessary functions are implemented in this board to establish communication and allow data management in higher layers under the MQTT protocol. A complementary monitoring interface was also developed using Node-RED. Several features were integrated within this interface, which includes an organized data displaying, real-time charts generation, and alarms that can be set and visualized based on the values of certain variables obtained through the developed gateway. The prototype and interface were tested to validate their operation and performance. The gateway is based on low-cost technologies and allows the integration of multiple protocols and industrial devices for possible future projects.

Keywords: Gateway · Industrial · IoT · MQTT · Node-RED

1 Introduction

In the last years, Industry 4.0 has emerged as a concept that refers to the intelligent networking of machines and processes for the industry with the help of information and communication technology. As result, many applications for using these intelligent networks in the industry have emerged too. For example, flexible production, customer-oriented solutions, optimized logistics, data use for optimization analyzing, predictive maintenance, resource-efficient circular economy, among others have been proposed [1]. But, Industry 4.0 paradigm is based on a set of enabling technologies, one of these technologies is the Internet of Things (IoT). The IoT has been defined in Recommendation ITU-T Y.2060 (06/2012) as a global infrastructure for the information society, enabling advanced services by interconnecting (physical and virtual) things based on existing and evolving interoperable information and communication technologies [2].

M. Botto-Tobar et al. (Eds.): *Latest Advances in Electrical Engineering, and Electronics*, LNEE 933, pp. 61–75, 2022.
https://doi.org/10.1007/978-3-031-08942-8_5

When IoT is implemented in industrial applications, it is defined as the industrial internet of things (IIoT). IIoT allows to connect in a network to devices, applications and even people that are involved in the execution of an industrial process. As result, several benefits could be achieved among which are [3]:

- Increase efficiency and productivity
- Create new business opportunities
- Strengthen worker safety
- Boost the product innovation process
- Reduces the cost of assets during their life cycle
- Improves understanding of consumer demands

All of these benefits can be real, but like all technological implementation requires an initial economic investment that not all industries are able to carry out. Such is the case of small and medium-sized industries. Therefore, it is important the availability of low-cost technology that can be accessed by industries with low purchasing power and what will allow them to access benefits of current technological advances. In this context, the present work proposes to begin work on a low-cost IIoT gateway prototype that allows the integration of process data into a virtual environment for small-scale industries.

This work is organized as follows. Section 2 presents a review of background works. Then, the selected protocols and technologies are briefly studied in Sect. 3. The requirements of the proposed prototype are exposed in Sect. 4. In Sect. 5 the procedure for implementing the proposed system is detailed. In Sect. 6 tests and their results are presented, and finally, the conclusions are drawn in Sect. 7.

2 Background

At the center of the IIoT is the gateway that provides connectivity between things and the cloud allowing to move data into the cloud as well as send control commands from the cloud to things. Gateway also allows compatibility between several devices although these use different protocols, because gateway is enabled to use different field protocols according to every field device needs [4].

In recent years commercial IIoT gateways such as Simatic IOT2000 gateway series as well as some prototypes have been developed. In the second case, many works have been carried out in order to develop and implement such gateways as mechanisms for implementing data acquisition systems specially for small-scale industrial plants. Although all works aim to provide an unified data structure for the industrial cloud and industrial big data [5], these proposed solutions are based on a variety of hardware, software, and technologies. For example, [6] presents a practical implementation based on a Raspberry Pi that allows data exchange between the user and a sensor network implemented using an Arduino. [7] proposes an IIoT gateway that enables users to supervise and control the plant through an online dashboard; this gateway is based on a Raspberry Pi 3 with its software developed in Python language and a server hosted in the IBM platform.

In Ecuador, several applications of the IIoT have been developed such as wireless sensor networks that use the emerging LoRa/LoRaWAN technologies as presented in [8], small-scale monitoring systems based on the Simatic IOT2000 gateway series [9], and even a monitoring architecture based on the low-cost single-board computer BeagleBone Black [10], in this architecture, the system allows data acquisition from an S7-1200 PLC and manages the MQTT protocol to communicate data to upper layers. Although this gateway solution seems promising the acquisition of such SBCs is still pretty difficult in the country.

In this context, the present work develops an IIoT gateway prototype based on an Arduino board that allows the integration of process data into a virtual environment for small-scale applications. Although multiple authors have proposed gateways based on Arduino, the solutions are oriented towards home automation [11,12], no industrial-oriented based on Arduino low-cost solutions could be found.

3 Selected Protocols and Technologies

According to a study conducted by HMS Networks in 2020 [13], current trend in industry is the use of industrial protocols such as Ethernet/IP, Profinet, Modbus TCP, EtherCAT, among others, which are based on Ethernet. However, legacy technologies based on field buses such as Modbus RTU, Profibus DP or DeviceNet are still needed to be integrated with the framework of new information technologies. For this reason, it is necessary to develop equipment that allows these devices to be integrated into a connectivity standard developed for IIoT. In this context, the field protocol selected for this work is Modbus RTU because it is one of the most widespread protocol in industrial applications. In the other hand, MQTT protocol is a standard messaging protocol, which has become the most used standard for IoT applications (as mentioned in [14]), especially for remote device integration. With this background, in the following section, we describe the operation of the previously stated communication protocols as well as technologies proposed for the development of our proposed system.

3.1 Modbus RTU Protocol

Modbus is an industry-standard protocol that enables the exchange of data between industrial control devices over a digital communication network. Modbus is one of the most widely used protocols in the industry since it is an open protocol that is versatile and easy to implement in the different levels of process automation. The protocol defines the structure of the messages and the mechanism by which the devices request information and generate responses.

Modbus RTU is one of the many variants of the Modbus protocol over a serial interface as RS-232 or RS-485. In where, devices access to interface using the master-slave technique, in which the master device initializes the transactions with every slave devices in a process known as "request". The active slave initiates the response to the request by performing a task and/or sending data to the master (see Fig. 1).

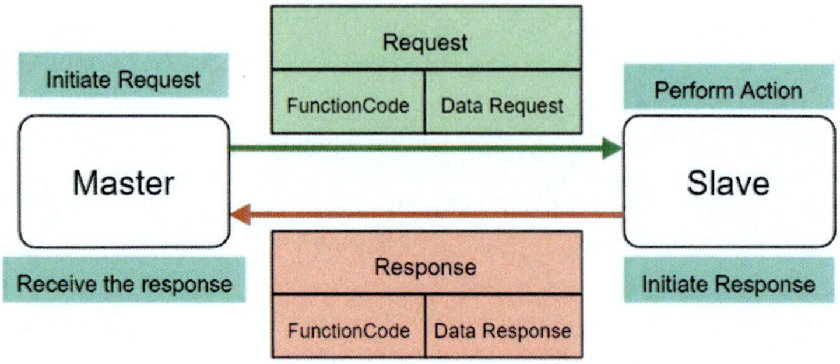

Fig. 1. Master-slave technique

3.2 MQTT Protocol

MQTT Protocol is a standard messaging protocol commonly used for internet of things applications [14]. It is based on a Publish/Subscribe architecture where the client that sends the message ("Publisher") is decoupled from the clients that receive the message ("Subscribers"), this means that the publishers and subscribers do not interact directly with each other, the connection is established by an additional component called Broker (Server) whose job is to filter incoming messages and distribute them to the subscribers who need them [15].

In Fig. 2 communication between three clients under the MQTT protocol is illustrated. This network is integrated by a single "Publisher" client with an embedded system capable of accessing a computer network, and two "Subscriber" clients. The "Publisher" client sends the value of a process variable (e.g. temperature) under a defined topic. Then, the "Broker" filters the message under that topic and delivers it to the "Subscriber" clients. Clients receive the process variable value under the topics to which are subscribed. If any client is not subscribed to the topic, the Broker will not deliver the message. This filtering method is known as "topic-based filtering".

3.3 Arduino Platform

Arduino is an open-source platform for the development of electronic prototypes using low-cost embedded boards based on microcontrollers that can be programmed through the Arduino integrated development environment (IDE) [16]. One of the most important features of this platform is the free access to several users contributions and collaborative projects that allow users to develop applications simply by using libraries to extend the board's functionalities with expansion modules, communications modules, storage devices, etc.

3.4 Node-RED

The Node-RED project is an open-source and free access software that allows devices to be integrated into Web services and other software. This has evolved

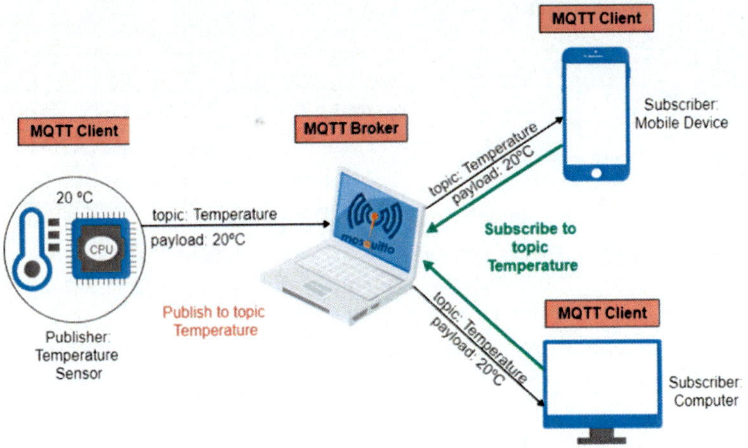

Fig. 2. Example of the MQTT publish/subscribe model.

as a general-purpose tool for programming IoT applications [17]. Then, it can be employed as a tool that supports the use and development of systems applied in the IoT field. Node-RED is based on visual flow programming in which the information travels through several blocks with different functionalities called "nodes" that allow developers to process information or develop tasks with input and output parameters.

4 Prototype Implementation Requirements

The main objective of our proposed system are to enable communication of industrial field data such as analog or digital inputs/outputs, data in Modbus RTU devices with serial communications, and low-power sensors. The purpose is to integrate the acquired data in the monitoring interface with two functionalities: (1) display of real-time trends and, (2) application for alarms management, both can be accessed from any device with internet connection.

The prototype must be developed using an embedded system capable of connecting to a local network and acting as an MQTT client. Data from inputs and outputs, as well as data from sensors and Modbus RTU device must be linked to publish/subscribe MQTT topics that will be managed by an MQTT Broker. Also, the use of a VPN (virtual private network) is required for remote access. It is important to highlight that the MQTT Broker, VPN service, and monitoring interface development tool are fundamental and may require a device with more computational capabilities than an embedded system, then They will be implemented in a personal computer.

Table 1 summarizes the requirements that have been taken into consideration for this project.

Table 1. Prototype requirements

Requirement	Description
Digital inputs	2 isolated 0–5 [V] digital inputs (DI0 and DI1)
Analog inputs	2 analog inputs of current (AI0) and voltage (AI1)
Digital output	4 Relay type digital outputs (D0-D3)
Analog output	1 Analog Output 0–10 [V] (AO0)
Industrial protocol: modbus RTU	Gather the data from contact (10000X) and input registers (30000X), and update the data of coil (00000X) and holding registers (40000X) of the station
I2C bus	Provide access to an I2C bus, in which devices can be added to enable the integration of other field devices, sensors or extend the number of inputs/outputs
MQTT protocol	Work as an MQTT Publish/Subscribe client
Monitoring interface	Represent the input values, in tables, gauges, indicators, and numeric displays. It should also need to have switches, and sliders to interact with the outputs. It has also been considered real-time trends, and an alarm system with e-mail notifications
Remote access	Ensure remote access via VPN

5 Proposed System Architecture

The main functionalities of the proposed prototype are drawn in the architecture shown in Fig. 3. These functionalities are acquisition of field data through the IIoT gateway both directly and indirectly. In the first case, data acquisition is accomplished through analog/digital inputs and outputs. In the second case, an I2C bus allows incorporating sensors and Modbus RTU field devices by means of I2C slave boards. For all cases, acquired data can be accessed by the monitoring interface. The components of the proposed architecture are reviewed in more detail below.

5.1 IIoT Arduino UNO Gateway (I2C Master) + Ethernet Shield

Its principal function is to act as an MQTT "Pub/Sub" client that publishes incoming field data and subscribes to command topics.

The function of the Arduino UNO is to process, manage and act upon the incoming bidirectional information flow to the board, in order to do so, the board needs to communicate by means of the serial peripheral interface (SPI) to the Ethernet module that is in charge of providing the TCP/IP functionalities to the Arduino, so the use of the Ethernet and SPI libraries is needed. To implement the MQTT client functionalities, the "PubSubClient" library is needed. Each one of the publishing messages has its own linked topic, the data from inputs, sensors, and registers of the Modbus RTU device are published in the main loop

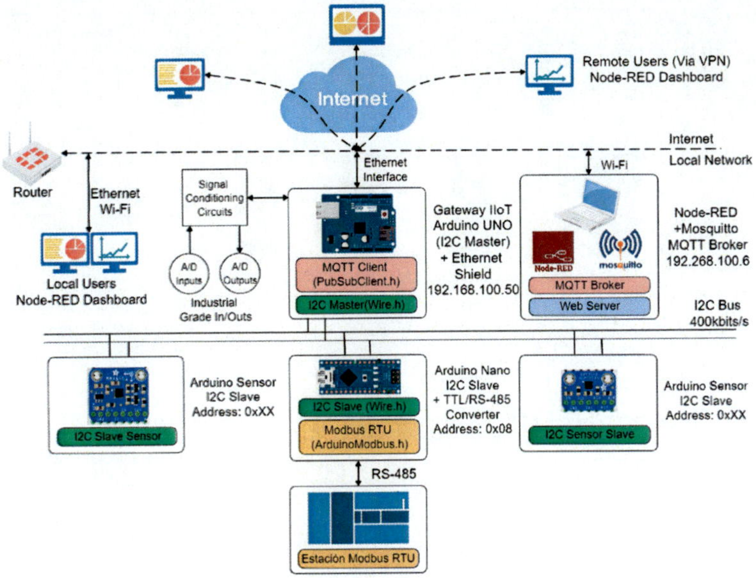

Fig. 3. Proposed system architecture

of the program in order to update the dashboard data frequently. In the case of the subscribed topics data, the device subscribes to the configured topics at the beginning of the program, then the library allows the reception of data by making a callback to a function where the data has to be processed each time there is incoming data linked to a filtered topic from the Node-RED's MQTT publishing nodes, this allows to activate its outputs or updating the Modbus RTU device's registers. To get the data from the Modbus RTU device and the sensors, the Arduino UNO also acts as an I2C bus master that allows collecting data of slave I2C sensors and an Arduino Nano I2C Slave that is responsible to manage the communication with the Modbus RTU device. It should also be noted that the use of the public I2C library Wire is necessary. At the beginning of the program, the system configures the I2C transmission speed at 400 kbits/s and initializes the I2C sensors. Then the incoming data from slaves is read and outgoing data from the master updates the registers of the slaves in the program's main loop by sending and receiving data through the I2C bus.

The use of the I2C bus for communication with other systems may allow the integration of even more devices that could get information of other types of industrial devices with different types of protocols or interfaces, different sensors or it might allow extending the number of available inputs and outputs. Allowing the system to become a modular device that can be a topic for future works. The number of devices might be limited by the number of I2C addresses that the master can manage, in the case of Arduino the slaves could use the addresses from 8 to 127 which means that the system could integrate up to 119 devices at the maximum speed of transmission is 400 kbits/s. The decoupling or failure

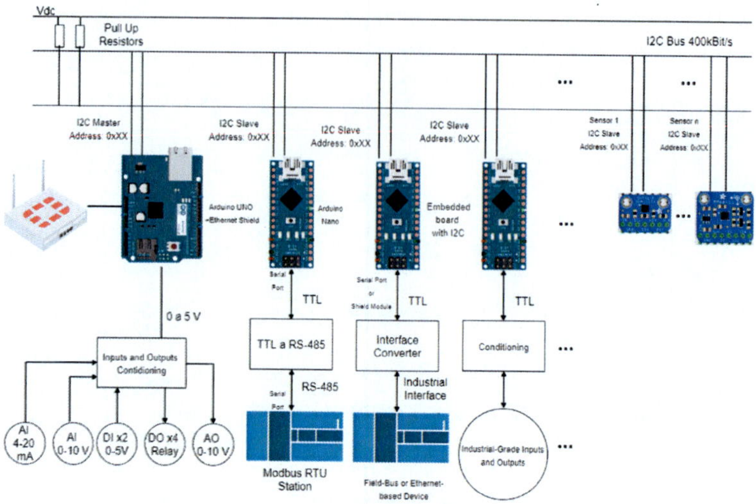

Fig. 4. I2C bus

of one of the slaves would not compromise the functionalities of the gateway, neither the functionality of the other slaves. The number of topics might be limited by the system memory. The I2C bus is presented in Fig. 4.

5.2 Arduino Nano (I2C Slave) + TTL/RS-485 Converter

This component (Shown in in Fig. 4) is based on an Arduino Nano board and allows to acquire information, via serial communications (through TTL/RS-485 conversion module) of the Modbus contact and input registers, and update the coil and holding registers of the Modbus field device. The upstream communication is made possible by means of the I2C bus to the Arduino UNO master in charge of managing the MQTT client topics of the Modbus registers.

5.3 I2C Slave Sensors

These sensors are used to gather data from physical variables, the chosen sensors are the accelerometer and gyroscope sensor MPU6050 and the digital pressure and temperature sensor BMP180, both of them are based on low power, low voltage processing units that have an I2C interface. It should be noted that the Arduino UNO master is in charge of initializing the sensors and processing incoming values before publishing them as MQTT messages.

5.4 Node-RED + Mosquitto MQTT Broker

These two components are implemented in a conventional computer that is in the same network as the developed IIoT gateway as shown in Fig. 3.

The MQTT Broker chosen is the lightweight open-source Mosquitto MQTT Broker, which is in charge of managing the communications of the messages by filtering them according to publication and subscription topics.

The second component is the Node-RED tool, which allows the generation of the monitoring interface by programming the flows. In order to do so Publish/Subscribe client nodes are needed, and other utility and contribution nodes that allow the integration of visualization widgets.

5.5 Inputs and Outputs Conditioning

A conditioning circuitry is needed in order to adapt the electrical industrial-type inputs and outputs of the gateway that guaranties isolation and adequate voltage levels to the Arduino UNO board.

The digital inputs must be optoisolated and the analog inputs must be adapted to the channel's voltage levels. In the same way, the digital outputs must be optoisolated and must activate relays. The analog output must be generated by a D/A converter managed by the board.

5.6 Router

It is a device that allows establishing the local network and enables access to the internet for the connected devices.

5.7 Users

Local Users: Computing devices capable of accessing the local network and interacting with the monitoring interface by employing an Ethernet or WiFi connection to the router.

Remote Users (Via VPN): Users that are capable of accessing the local network and interacting with the monitoring interface through a virtual private network tunnel managed by a VPN service. The VPN service is implemented using Hamachi by LogMeIn [18].

5.8 Modbus RTU Station

Modbus RTU station represents a device such as PLCs or industrial devices that can exchange information through the Modbus RTU protocol.

6 Performed Tests and Results

The performance of each of the prototype's functions was evaluated through mechanisms that were used to emulate data from a practical industrial process. In the case of data acquisition from the inputs and outputs, they are validated through electrical signals and measured values of pressure, altitude, and temperature acquired by a BMP180 sensor and the measured values of linear accelerations and angular speeds acquired by an MPU6050 sensor. In order to carry out the test of Modbus communication, the free Modbus simulator ModRSsim2 [19] is used to allow the simulation of Modbus RTU slaves and TCP/IP servers.

Since the monitoring interface developed in Node-RED has three main windows. Then, tests of prototype functions are described according to interface displays of acquired data.

6.1 Devices

In this screen, three tabs are used to display each one of the different dashboards of the field devices. In each dashboard, a summary table of all the variables is presented in order to provide additional information related to the process.

Analog and Digital Inputs and Outputs. In this dashboard (shown in Fig. 5) the values of the analog inputs and digital inputs are displayed in gauges and indicator lights respectively. There are also switches and sliders that allow activating the outputs of the device.

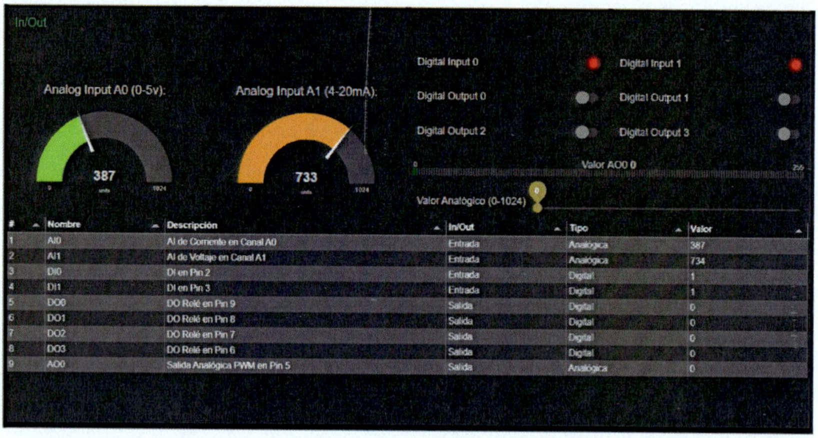

Fig. 5. Inputs and outputs dashboard

Modbus RTU. This dashboard is used to represent the values of 8 contact registers that are represented by indicator lights, 4 input registers that are represented in gauges, 8 coil registers which states can be changed by interacting with switches, and 4 holding registers that can be changed by moving 4 sliders. The dashboard is presented in Fig. 6 with the registers of the simulated station.

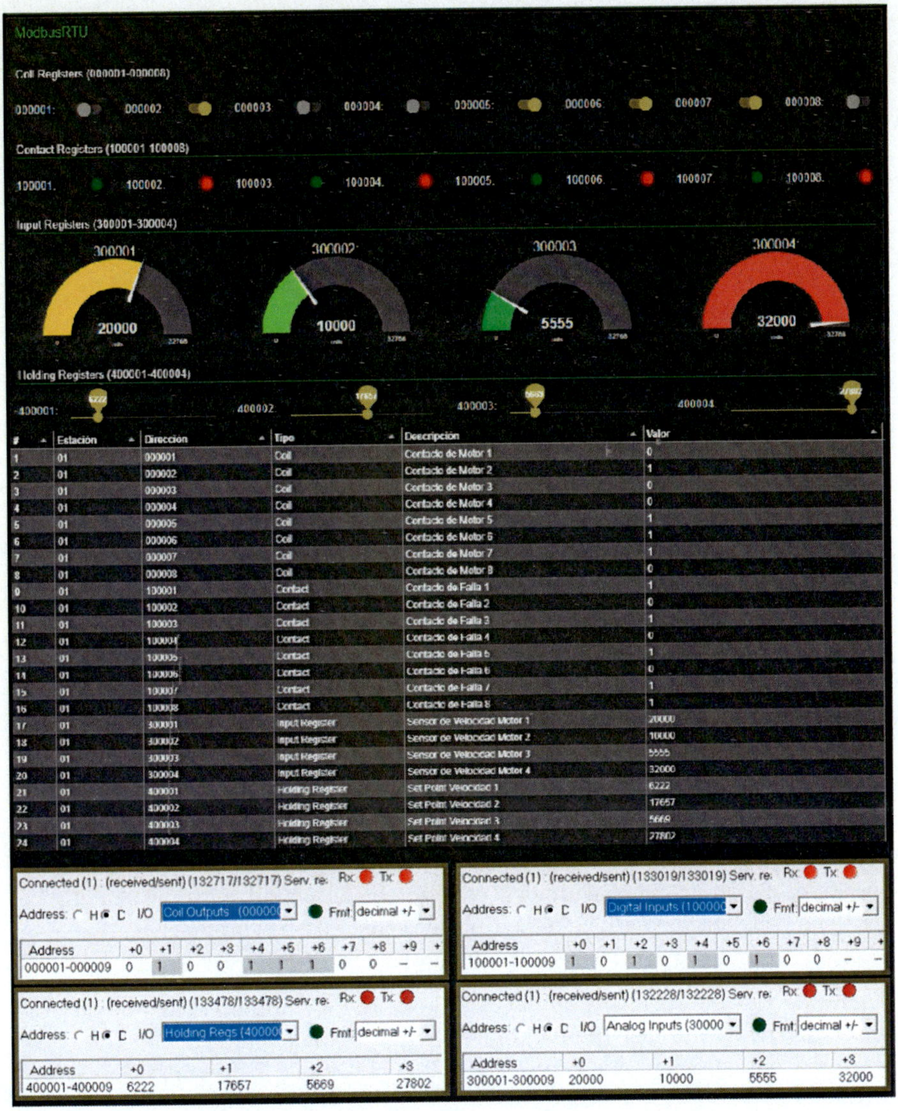

Fig. 6. Modbus dashboard and modbus registers

Sensors. This dashboard (Fig. 7) presents the measured values of pressure, altitude, and temperature acquired by the BMP180 sensor and the measured values of linear accelerations and angular speeds acquired by the MPU6050 sensor.

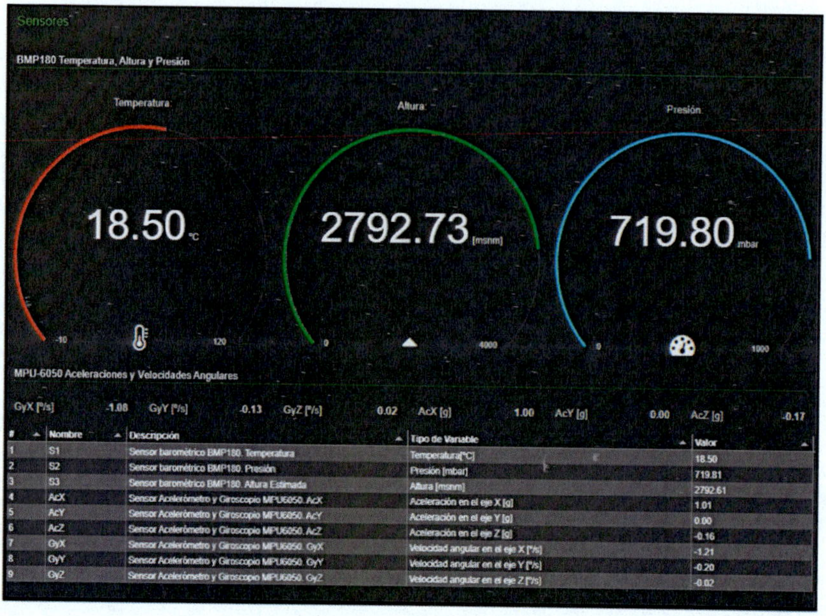

Fig. 7. Sensors dashboard

6.2 Alarms

This screen has two tabs that allow the display of two dashboards as proposed in the previous sections, the Node-RED programming tool is used to generate an alarm configuration dashboard and an alarm register dashboard, the alarm consists of a pop-up message and if configured, an e-mail must be sent to the user. The dashboards and the e-mail notification are shown in Fig. 8.

Alarm Register. In this dashboard is presented a summary table of the configured alarms (including the enabling, logging, e-mail sending, and the alarm threshold configurations) and a register table of the last 20 generated alarms.

Alarms Configurations. This dashboard allows the user to enable an alarm and configure its logging in the previous dashboard, automatic notifications via e-mail, and the alarm's thresholds.

As shown in Fig. 8, the notification is automatically sent when the digital input alarm is triggered.

Fig. 8. Alarm register, alarm configuration and e-mail notification

6.3 Real-Time Trends

In this screen, many of the digital and numeric variables obtained from the field by the gateway are represented in real-time trend graphs with a time interval of 15 s. In Fig. 9 is presented the real-time trend of the Modbus Input registers which were modified from the simulated Modbus station.

Fig. 9. Real-time trends

7 Conclusions and Future Works

This paper has presented a proposal for IIoT gateway development by integrating the most popular open-access tools, protocols, and technologies available. The proposed prototype consists of an MQTT Publish/Subscribe client that allows bidirectional interaction between field devices and users. Also, the computer system provides access to local and remote users to the monitoring interface, integrating visualization and alarm functionalities.

The proposed prototype is based on a low-cost embedded board, making it suitable for industries with low purchasing power. Also, since the gateway is based on open-source technologies that provide its modular characteristics, it may allow a future integration to an Industry 4.0 framework of the multiple field buses or industrial protocols based on Ethernet.

Among future jobs can be noted that in the case of gateway integration into the industrial environment, hardware improvements are required in order to guaranty its functionality according to industrial certifications and standards.

References

1. The website of the Federal Ministry for Economic Affairs and Energy of Germany (2021). https://www.plattform-i40.de/PI40/Navigation/EN/Industrie40/WhatIsIndustrie40/what-is-industrie40.html, Accessed 10 Sep 2021
2. The website of the ITU. https://www.itu.int, Accessed 10 Sep 2021
3. Alessandrini, G.: Internet Industrial de las Cosas (IIoT). Instituto de Calidad Industrial (INCALIN), Universidad Nacional de San Martin, Buenos Aires (2021)

4. Munirathinam, S.: Chapter six - industry 4.0: industrial internet of things (IIOT). In: Raj, P., Evangeline, P. (Eds.), The Digital Twin Paradigm for Smarter Systems and Environments: The Industry Use Cases. Advances in Computers, vol. 117, pp. 129–164. Elsevier, Amsterdam (2020). https://doi.org/10.1016/bs.adcom.2019.10.010
5. Shimei, L. Jianhong, Z., Enfeng, L. Gang, H.: Design of industrial internet of things gateway with multi-source data processing. In: International Conference on Computer Engineering and Application (ICCEA), pp. 232–236 (2020). https://ieeexplore.ieee.org/document/9103715
6. Glória, A., Cercas, F., Souto, N.: Design and implementation of an IoT gateway to create smart environments. Procedia Comput. Sci. **109**, 568–575 (2017). https://doi.org/10.1016/j.procs.2017.05.343
7. Tietz, F. Brandão, D. Alves, L. F.: Development of an internet of things gateway applied to a multitask industrial plant. In: 13th IEEE International Conference on Industry Applications (INDUSCON), pp. 917–923 (2018). https://doi.org/10.1109/INDUSCON.2018.8627273
8. Heredia, A., Lucero, P., Astudillo, F., Vázquez, A.: Design and implementation of wireless sensor network with LoRa technology for industrial monitoring. Latin-Am. J. Comput. (LAJC) **7**, 50–60 (2020)
9. Vargas, J.: Diseño e implementación de una red de controladores usando el gateway Simatic IOT2040 y un dashboard de acceso local y remoto (2021). https://bibdigital.epn.edu.ec/handle/15000/21423
10. Borja, E.: Diseño de una arquitectura usando el protocolo message queue telemetry transport (MQTT) sobre plataformas de bajo coste, para monitorización de procesos industriales (2020). http://dspace.espoch.edu.ec/handle/123456789/14099
11. Grygoruk, A., Legierski, J.: IoT gateway - implementation proposal based on Arduino board, In: Federated Conference on Computer Science and Information Systems (FedCSIS), pp. 1011–1014 (2016). https://doi.org/10.15439/2016f283
12. Kodali, R.K., Anjum, A.: IoT based home automation using node-RED. In: Second International Conference on Green Computing and Internet of Things (ICGCIoT), pp. 386–390 (2018). https://doi.org/10.1109/ICGCIoT.2018.8753085
13. The website of HMS networks. https://www.hms-networks.com/news-and-insights/news-from-hms/2020/05/29/industrial-network-market-shares-2020-according-to-hms-networks, Accessed 30 Aug 2021
14. The website of the MQTT Org (2021). https://mqtt.org, Accessed 19 Aug 2021
15. MQTT and MQTT 5 Essentials. 1st edn. HiveMQ GmbH, Ergoldinger Str. 2A 84030, Landshut Germany (2020)
16. The website of Arduino. https://www.arduino.cc/en/Guide/Introduction, Accessed 30 Aug 2021
17. Node-RED Programming Guide. http://noderedguide.com, Accessed 30 Aug 2021
18. The Hamachi Website by LogMeIn. https://www.vpn.net, Accessed 30 Aug 2021
19. Modbus Technical Resources. https://modbus.org/tech.php, Accessed 30 Aug 2021

The Quadruple-Tank Process: A Comparison Among Advanced Control Techniques

Jhostin Cisneros[1], Cinthya Orbe[1], Carol Paucar[1], Francisco Toapanta[1], and Oscar Camacho[2](✉)

[1] Facultad de Ingeniería Eléctrica y Electrónica, Escuela Politécnica Nacional, Quito, Ecuador
{jhostin.cisneros,cinthya.orbe,carol.paucar, francisco.toapanta}@epn.edu.ec

[2] Colegio de Ciencias e Ingenierías "El Politécnico", Universidad San Francisco de Quito USFQ, Quito 170157, Ecuador
ocamacho@usfq.edu.ec

Abstract. A system composed of four tanks is presented for which it's shown its mathematical model and the obtaining of different types of controllers considering that it has a multivariable system, so it is important to define the interaction between the input and output variables to choose the possible transfer functions on which to base the design of the controllers, or to design the decouplers and obtain the design base transfer functions with them. The results of the implementation of different designed controllers are going to be presented and their performance will be evaluated through indexes and transitory parameters such as its overshoot and establishment time. Based on all of this, the conclusion of this work are going to be presented as the better option to implement the control on this system and some considerations that have been shown in said results.

Keywords: Multivariable process · Control techniques

1 Introduction

Since its first appearance in the literature, the Quadruple Tank process, proposed by Johansson in 2000, has been cataloged as a benchmark for design control schemes. Due to it is a nonlinear multivariable system, which due to interactions can change the position of the zeros of the linearized model between the right and left side of the half-plane, together with its relatively simple construction, makes it a helpful model for teaching control [7]. This system consists of four interconnected tanks so that the outflows from the upper tanks are discharged into the lower tanks. The flow is supplied to the tanks through two circuits with two identical pumps. Each circuit has a three-way valve that distributes the flow provided between the upper and lower tanks. Specifically, the system has two inputs: the actuators of the three-way valves (V1, V2) and two outputs, which are the tank levels (h1, h2), as shown Fig. 1.

A wide variety of control systems have been proposed on this system, ranging from the schemes presented in [8] consisting of decentralized PID control and PID-Fuzzy

M. Botto-Tobar et al. (Eds.): *Latest Advances in Electrical Engineering, and Electronics*, LNEE 933, pp. 76–88, 2022.
https://doi.org/10.1007/978-3-031-08942-8_6

control. Reference model-based schemes such as the one shown in [9] use the MIT rule to define the adaptive parameters. Robust SMC-based schemes such as the one in [10] that implement an SMC control based on a Linear Quadratic Regulator with integral action as the sliding surface. Linear algebra-based schemes such as those of [2, 3] and [11] are characterized by avoiding the use of decouplers thanks to the action of the sacrificed variable.

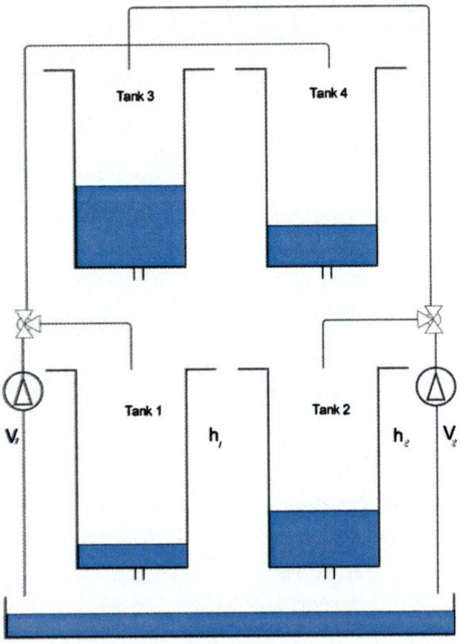

Fig. 1. Four tanks system

The objective of this work is to make a comparison of different advanced control schemes on the quadruple tank process. For this, the mathematical modeling of the plant will be performed in Sect. 2. In Sect. 3, the corresponding decouplers will be obtained following the Bristol method to implement the following control schemes finally: traditional PID tuned by Dahlin, SMC according to [1, 5], MRAC, and Linear Algebra according to the procedure of [4, 6], in addition to using a variant of this technique that adapts to the use of the previously designed decouplers. The result graphics and performance indices are presented in Sect. 4. Finally, Sect. 5 presents the conclusions of the work done after a discussion of the results.

2 Materials

The dynamic of the system was defined in the introduction, so that this section will present its parameters and values to the minimum phase operation [5].

Where:

$$\boldsymbol{A_i}\text{: } Cross\ section\ of\ the\ tank\ i.\ (A_1, A_2, A_3, A_4)$$

$$\boldsymbol{a_i}\text{: } cross\ section\ of\ tank\ outlet\ duct\ A_i$$

$\boldsymbol{h_i}$: *liquid level in the tank i* (h_1, h_2, h_3, h_4)
$\boldsymbol{h_{i0}}$: *nominal liquid level in the tank i* (h_{10}, h_{20}, h_{30}, h_{40})
$\boldsymbol{k_c}$: *level sensor gain*
$\boldsymbol{g}$: *gravity acceleration*
$\boldsymbol{k_1}$, $\boldsymbol{k_2}$: *pump gain* 1 *and* 2
$\boldsymbol{v_1}$, $\boldsymbol{v_2}$: *pump input voltage* 1 *and* 2
$\boldsymbol{v_{10}}$, $\boldsymbol{v_{20}}$: *nominal pump input voltage* 1 *and* 2
$\boldsymbol{\gamma_1}$, $\boldsymbol{\gamma_2}$: *flow proportion to the tank* 1 *and* 2

Table 1. Parameters of the model

Parameter [unit]	Value
$A_1; A_3\left[cm^2\right]$	28
$A_2, A_4\left[cm^2\right]$	32
$a_1, a_3\left[cm^2\right]$	0.071
$a_2, a_4\left[cm^2\right]$	0.057
$g\left[cm/s^2\right]$	981
$h_{10}; h_{20}[cm]$	12.263; 12.783
$h_{30}; h_{40}[cm]$	1.63394; 1.40904
$\gamma_1; \gamma_2$	0.7; 0.6
$v_{10}; v_{20}[V]$	3; 3
$k_1; k_2\left[\frac{cm^3}{Vs}\right]$	3.33; 3.35

The model equations are as follows [5] (Table 1):

$$\frac{dh_1}{dt} = -\frac{a_1}{A_1}\sqrt{2gh_1} + \frac{a_3}{A_1}\sqrt{2gh_3} + \frac{\gamma_1 k_1}{A_1}v_1 \tag{1}$$

$$\frac{dh_2}{dt} = -\frac{a_2}{A_2}\sqrt{2gh_2} + \frac{a_4}{A_2}\sqrt{2gh_4} + \frac{\gamma_2 k_2}{A_2}v_2 \tag{2}$$

$$\frac{dh_3}{dt} = -\frac{a_3}{A_3}\sqrt{2gh_3} + \frac{(1-\gamma_2)k_2}{A_3}v_2 \tag{3}$$

$$\frac{dh_4}{dt} = -\frac{a_4}{A_4}\sqrt{2gh_4} + \frac{(1-\gamma_1)k_1}{A_4}v_1 \tag{4}$$

Relative Gain Matrix utilizing approximate linearization

$$T_i = \frac{A_i}{a_i}\sqrt{\frac{2h_{i0}}{g}} \quad i = 1, 2, 3, \tag{5}$$

$$\frac{dh}{dt} = \begin{bmatrix} -\frac{1}{T_1} & 0 & \frac{A_3}{T_3A_1} & 0 \\ 0 & -\frac{1}{T_2} & 0 & \frac{A_4}{T_4A_2} \\ 0\,0 & & -\frac{1}{T_3} & 0 \\ 0\,0 & & 0 & -\frac{1}{T_4} \end{bmatrix} \begin{bmatrix} h_1 \\ h_2 \\ h_3 \\ h_4 \end{bmatrix} + \begin{bmatrix} \frac{\gamma_1 k_1}{A_1} & 0 \\ 0 & \frac{\gamma_2 k_2}{A_2} \\ 0 & \frac{(1-\gamma_2)k_2}{A_3} \\ \frac{(1-\gamma_1)k_1}{A_4} & 0 \end{bmatrix} \begin{bmatrix} V_1 \\ V_2 \end{bmatrix} \tag{6}$$

$$y = \begin{bmatrix} k_c & 0 & 0\,0 \\ 0 & k_c & 0\,0 \end{bmatrix} \begin{bmatrix} h_1 \\ h_2 \\ h_3 \\ h_4 \end{bmatrix} \tag{7}$$

Equations can also be expressed in transfer functions [1, 5]:

$$h_1(s) = \frac{A_3T_1}{T_3A_1(T_1S + 1)} h_3(s) + \frac{\gamma_1 k_1 T_1}{A_1(T_1S + 1)} u_1(s) \tag{8}$$

$$h_2(s) = \frac{A_4T_2}{T_4A_2(T_2S + 1)} h_4(s) + \frac{\gamma_2 k_2 T_2}{A_2(T_2S + 1)} u_2(s) \tag{9}$$

$$h_3(s) = \frac{(1 - \gamma_2)k_2T_3}{A_3(T_3S + 1)} u_2(s) \tag{10}$$

$$h_4(s) = \frac{(1 - \gamma_1)k_1T_4}{A_4(T_4S + 1)} u_1(s) \tag{11}$$

When applying the formula $G(s) = C(SI - A)^{-1}B + D$

$$G(s) = \begin{bmatrix} \frac{2.595}{62.35s+1} & \frac{1.492}{(62.35s+1)(22.76s+1)} \\ \frac{1.414}{(90.63s+1)(30.089s+1)} & \frac{2.84}{90.63s+1} \end{bmatrix} \tag{12}$$

steady-state gain matrix:

$$K = \begin{pmatrix} 2.595 & 1.492 \\ 1.414 & 2.84 \end{pmatrix} \tag{13}$$

Using the Schur product, the RGA matrix is determined:

$$\Lambda = K\left(K^{-1}\right)^T = \begin{bmatrix} 1.401 & -0.401 \\ -0.401 & 1.401 \end{bmatrix} \tag{14}$$

With the above result, it is shown that only the function $G_{11}(s)$ *and* $G_{22}(s)$ should be controlled as the others give negative responses, to control these you can get unwanted results.

With this result, it is decided that it is better a couple of the input one with the output one and the input two with the output two.

2.1 Plant Model

$$G_{11}(s) = \frac{2.68\, e^{-0.4}}{65.4s + 1};\; G_{21}(s) = \frac{1.44\, e^{-24.9}}{104.08s + 1} \tag{15}$$

$$G_{12}(s) = \frac{1.52\, e^{-18.1}}{73.29s + 1};\; G_{22}(s) = \frac{2.94\, e^{-0.6}}{95.4s + 1} \tag{16}$$

Matrix-based on first order plus delay time models (FOPDT)

$$G(s) = \begin{pmatrix} \frac{2.68\, e^{-0.4}}{65.4s+1} & \frac{1.52\, e^{-18.1}}{73.29s+1} \\ \frac{1.44\, e^{-24.9}}{104.08s+1} & \frac{2.94\, e^{-0.6}}{95.4s+1} \end{pmatrix} \tag{17}$$

RGA matrix

$$\Lambda = \begin{pmatrix} 1.384 & -0.384 \\ -0.384 & 1.384 \end{pmatrix} \tag{18}$$

3 Advanced Control Designs

For the design of controllers, four classes were considered as the adaptive control, control by sliding modes, and linear algebra, in the latter it is done with two meters with the use of decouplers and without these:

$$D(s) = \begin{bmatrix} 1 & D_{12}(s) \\ D_{21}(s) & 1 \end{bmatrix} \tag{19}$$

$$D_{21}(s) = -\frac{0.57498}{22.761s + 1};\; D_{12}(s) = -\frac{0.49788}{30.089s + 1} \tag{20}$$

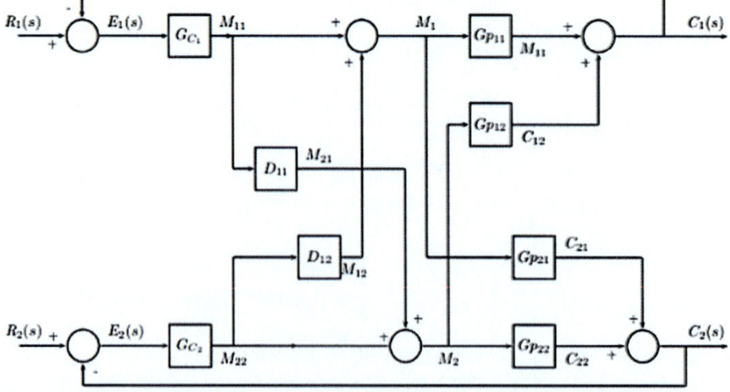

Fig. 2. Typic decoupling system 2 × 2 [5, 12]

3.1 PID Control

By Dahlin's tuning, it's possible to obtain the values for the PID controller; for that, it is necessary to use the FOPDT, based on the result of the RGA matrix, just must be controlled two models, the height of the first and second tank.

$$G_p(s) = \frac{kme^{-t_0}}{\tau s + 1}; k_p = \frac{1}{km} * \left(\frac{\tau}{t_0}\right); \tau_i = \tau; \tau_d = \frac{t_0}{2} \tag{21}$$

$$G_c(s) = k_p\left(1 + \frac{1}{\tau_i s} + \tau_d s\right) \tag{22}$$

$$G_{c11}(s) = 30.5\left(1 + \frac{1}{65.4s} + 0.2s\right) \tag{25}$$

$$G_{c22}(s) = 27.04\left(1 + \frac{1}{95.4s} + 0.3s\right) \tag{26}$$

3.2 Sliding Mode Control (SMC)

$$G_{11}(s) = \frac{2.68\, e^{-0.4}}{65.4s + 1}; G_{22}(s) = \frac{2.94e^{-0.6}}{95.4s + 1} \tag{25}$$

The SMC was determined, with the proposal of Camacho and Smith [5]

$$U_c(t) = \left(\frac{\tau * t_0}{K}\right)\left[\frac{X(t)}{\tau * t_0} + \lambda_0 e(t)\right] + K_D(t)\frac{S(t)}{|S(t)| + \delta} \tag{26}$$

$$S(t) = sign(K)\left[-\frac{dX(t)}{dt} + \lambda_1 e(t)\right] \tag{27}$$

$$\lambda_1 = \frac{t_0 + \tau}{t_0 \tau}$$

$$\lambda_0 = \frac{\lambda_1^2}{4} \tag{29}$$

Discontinuous part

$$K_D = \frac{0.51}{|K|}\left(\frac{\tau}{t_0}\right); \delta = 0.68 + 0.12|K|K_D\lambda_1 \tag{30}$$

3.3 Model Reference Adaptative Control (ADAPTATIVE)

This model has three main elements: Reference Model, Plant Model, and Adaptative Controller. The overall model looks like this (Fig. 3):

1. Reference model, in this case, it is necessary to use a closed-loop system. There can be used, the schematic that is shown in Fig. 2. Using the controller syntonized by Dalhin.

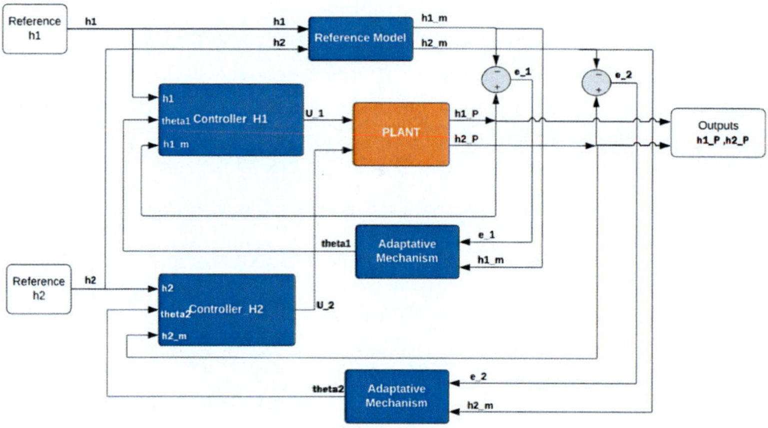

Fig. 3. Schematic of ADAPTATIVE [6]

2. Plant block has been made in the first steps of this document
3. In the Controller block are three signals: Reference, Signal output (Hx_P), and theta
4. Theta comes from the Adaptative block Mechanism, which multiplies the output from the Controller PID. Theta is equal to the product between Gamma which can be a value between 0–1, and the product of hx_m and the error between the signal of the plant model and the result of reference model. While the value is near 0, the response becomes fast; for this case, use a Gamma 0.05 for both controllers, which gives the best responses [6].

3.4 Linear Algebra with Decouplers (LA-D)

Based on the use of decouplers, you can determine the control law using Euler:

$$h_{n+1} = h_n + To\left(-\frac{1}{T_1}h_1 + \frac{A_3}{T_3A_1}h3 + \frac{r_1k_1}{A_1}V_1\right) \tag{31}$$

$$e_{n+1} = ke_n \tag{32}$$

$$h_{n+1ref} - h_{n+1} = ke_n \tag{33}$$

$$h_{n+1} = h_{n+1ref} - ke_n \tag{34}$$

$$V_1 = \frac{A_1}{K_1r_1}\left(\frac{h_{n+1ref} - ke_n - h_n}{To} + \frac{1}{T_1}h_1 - \frac{A_3}{T_3A_1}h3\right) \tag{35}$$

$$V_2 = \frac{A_2}{K_2r_2}\left(\frac{h_{n+2ref} - ke_n - h_n}{To} + \frac{1}{T_2}h_2 - \frac{A_4}{T_4A_2}h_4\right) \tag{36}$$

3.5 Linear Algebra Without Decouplers (LA)

The mathematical model of the tanks system can be expressed as follow [7].

$$\begin{bmatrix} \dot{h}_1 \\ \dot{h}_2 \\ \dot{h}_3 \\ \dot{h}_4 \end{bmatrix} = \begin{bmatrix} -\frac{a_1}{A1}\sqrt{2gh_1} + \frac{a_3}{A1}\sqrt{2gh_3} \\ -\frac{a_2}{A2}\sqrt{2gh_2} + \frac{a_4}{A2}\sqrt{2gh_4} \\ -\frac{a_3}{A3}\sqrt{2gh_3} \\ -\frac{a_4}{A4}\sqrt{2gh_4} \end{bmatrix} + \begin{bmatrix} \frac{\gamma_1 k_1}{A1}\ 0 \\ 0\ \frac{\gamma_2 k_2}{A2} \\ 0\ \frac{(1-\gamma_2)k_2}{A3} \\ \frac{(1-\gamma_1)k_1}{A4}\ 0 \end{bmatrix} \begin{bmatrix} v_1 \\ v_2 \end{bmatrix} \tag{37}$$

We can find the law control by using the Euler approximation to have:

$$\begin{bmatrix} \frac{\gamma_1 k_1}{A1}\ 0 \\ 0\ \frac{\gamma_2 k_2}{A2} \\ 0\ \frac{(1-\gamma_2)k_2}{A3} \\ \frac{(1-\gamma_1)k_1}{A4}\ 0 \end{bmatrix} \begin{bmatrix} v_1 \\ v_2 \end{bmatrix} = \begin{bmatrix} \frac{h_{1(n+1)}-h_{1(n)}}{T_0} + \frac{a_1}{A1}\sqrt{2gh_{1(n)}} - \frac{a_3}{A1}\sqrt{2gh_{3(n)}} \\ \frac{h_{2(n+1)}-h_{2(n)}}{T_0} + \frac{a_2}{A2}\sqrt{2gh_2} - \frac{a_4}{A2}\sqrt{2gh_4} \\ \frac{h_{3(n+1)}-h_{3(n)}}{T_0} + \frac{a_3}{A3}\sqrt{2gh_3} \\ \frac{h_{4(n+1)}-h_{4(n)}}{T_0} + \frac{a_4}{A4}\sqrt{2gh_4} \end{bmatrix} \tag{38}$$

Rewriting the equation, the system's state at time $n + 1$ can be determinate using numerical methods to calculate the control actions at time n based on the dynamic of the error.

$$\begin{bmatrix} \gamma_1 k_1\ 0 \\ 0\ \gamma_2 k_2 \\ 0\ (1-\gamma_2)k_2 \\ (1-\gamma_1)k_1\ 0 \end{bmatrix} \begin{bmatrix} v_1 \\ v_2 \end{bmatrix} = \begin{bmatrix} A1\frac{h_{1r(n+1)}-\lambda_1\left(h_{1r(n)}-h_{1(n)}\right)-h_{1(n)}}{T_0} + a_1\sqrt{2gh_{1(n)}} - a_3\sqrt{2gh_{3(n)}} \\ A2\frac{h_{2r(n+1)}-\lambda_2\left(h_{2r(n)}-h_{2(n)}\right)-h_{2(n)}}{T_0} + a_2\sqrt{2gh_2} - a_4\sqrt{2gh_4} \\ A3\frac{h_{3r(n+1)}-\lambda_3\left(h_{3r(n)}-h_{3(n)}\right)-h_{3(n)}}{T_0} + a_3\sqrt{2gh_3} \\ A4\frac{h_{4r(n+1)}-\lambda_4\left(h_{4r(n)}-h_{4(n)}\right)-h_{4(n)}}{T_0} + a_4\sqrt{2gh_4} \end{bmatrix} \tag{39}$$

Then an exact solution for the system is searched to satisfy the parallelism relation.

$$\frac{\gamma_1}{(1-\gamma_1)} = \frac{A1\frac{h_{1r(n+1)}-\lambda_1\left(h_{1r(n)}-h_{1(n)}\right)-h_{1(n)}}{T_0} + a_1\sqrt{2gh_{1(n)}} - a_3\sqrt{2gh_{3(n)}}}{A4\frac{h_{4r(n+1)}-\lambda_4\left(h_{4r(n)}-h_{4(n)}\right)-h_{4(n)}}{T_0} + a_4\sqrt{2gh_4}} \tag{40}$$

$$\frac{\gamma_2}{(1-\gamma_2)} = \frac{A2\frac{h_{2r(n+1)}-\lambda_2\left(h_{2r(n)}-h_{2(n)}\right)-h_{2(n)}}{T_0} + a_2\sqrt{2gh_2} - a_4\sqrt{2gh_4}}{A3\frac{h_{3r(n+1)}-\lambda_3\left(h_{4r(n)}-h_{3(n)}\right)-h_{3(n)}}{T_0} + a_3\sqrt{2gh_3}} \tag{41}$$

These solutions represent the conditions for the system to have an exact solution, and it gives the reference to h3 and h4, being them the sacrificed variables.

$$h_{3r(n+1)} = To[\frac{1}{A3}(\frac{(1-\gamma_2)}{\gamma_2}(A2\frac{h_{2r(n+1)} - \lambda_2\left(h_{2r(n)} - h_{2(n)}\right) - h_{2(n)}}{T_0} + a_2\sqrt{2gh_2} - a_4\sqrt{2gh_4}) - a_3\sqrt{2gh_3}) + \lambda_3\left(h_{3r(n)} - h_{3(n)}\right) + h_{3(n)}] \tag{42}$$

$$h_{4r(n+1)} = To[\frac{1}{A4}(\frac{(1-\gamma_1)}{\gamma_1}(A1\frac{h_{1r(n+1)} - \lambda_1(h_{1r(n)} - h_{1(n)}) - h_{1(n)}}{T_0} + \ldots$$
$$+ a_1\sqrt{2gh_{1(n)}} - a_3\sqrt{2gh_{3(n)}}) - a_4\sqrt{2gh_4}) + \lambda_4(h_{4r(n)} - h_{4(n)})$$
$$+ h_{4(n)}] \tag{43}$$

By this process, the law control is obtained using the transposed of the A matrix, which is the matrix that accompanies the manipulated variables.

$$\left(\gamma_1^2 + (1-\gamma_1)\right)^2)v_1$$
$$= \gamma_1(A1\frac{h_{1r(n+1)} - \lambda_1(h_{1r(n)} - h_{1(n)}) - h_{1(n)}}{T_0} + a_1\sqrt{2gh_{1(n)}}$$
$$- a_3\sqrt{2gh_{3(n)}})$$
$$+ (1-\gamma_1)(A4\frac{h_{4r(n+1)} - \lambda_4(h_{4r(n)} - h_{4(n)}) - h_{4(n)}}{T_0} + a_4\sqrt{2gh_4}) \tag{44}$$

$$\left(\gamma_2^2 + (1-\gamma_2)\right)^2)v_2$$
$$= \gamma_2(A2\frac{h_{2r(n+1)} - \lambda_2(h_{2r(n)} - h_{2(n)}) - h_{2(n)}}{T_0} + a_2\sqrt{2gh_2}$$
$$- a_4\sqrt{2gh_4})$$
$$+ (1-\gamma_2)(A3\frac{h_{3r(n+1)} - \lambda_3(h_{3r(n)} - h_{3(n)}) - h_{3(n)}}{T_0}$$
$$+ a_3\sqrt{2gh_3}) \tag{45}$$

4 Simulation Results

Once the control law mathematical expression is established, tuning their parameters is necessary to improve its response. Additionally, to measure its performance, Adaptative, SMC, LA-D, LA responses will be compared with a traditional method for this application PID.

4.1 Set-Point Step Change Response

The variation considered in the reference goes from the initial value of h1 = 6.13 [cm], h2 = 6.39 [cm]. The results obtained are those shown in Fig. 4 and Fig. 5.

4.2 Disturbance is Rejection

The curves obtained from the disturbances applied to the Quadruple-Tank Process are those shown in Fig. 6 to Fig. 7. Figure 8 shows the performance indicators to compare controllers under disturbances.

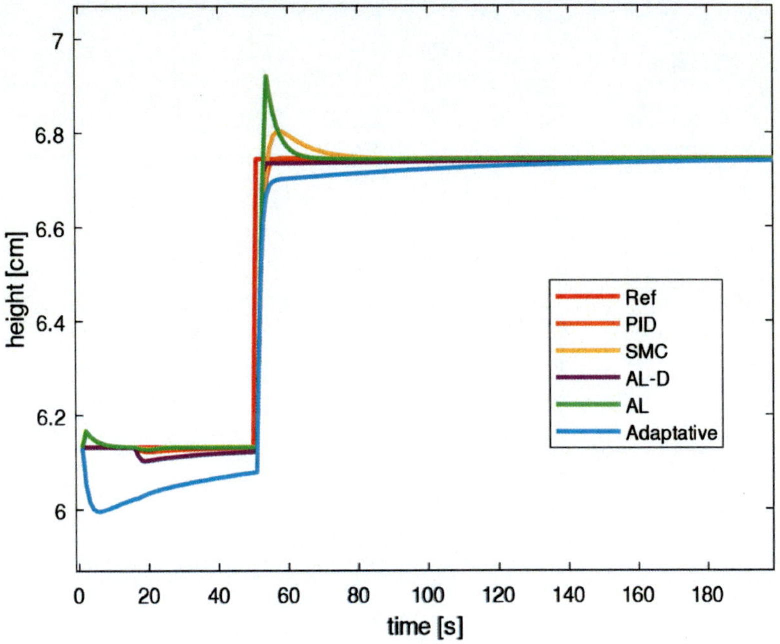

Fig. 4. Step response H1

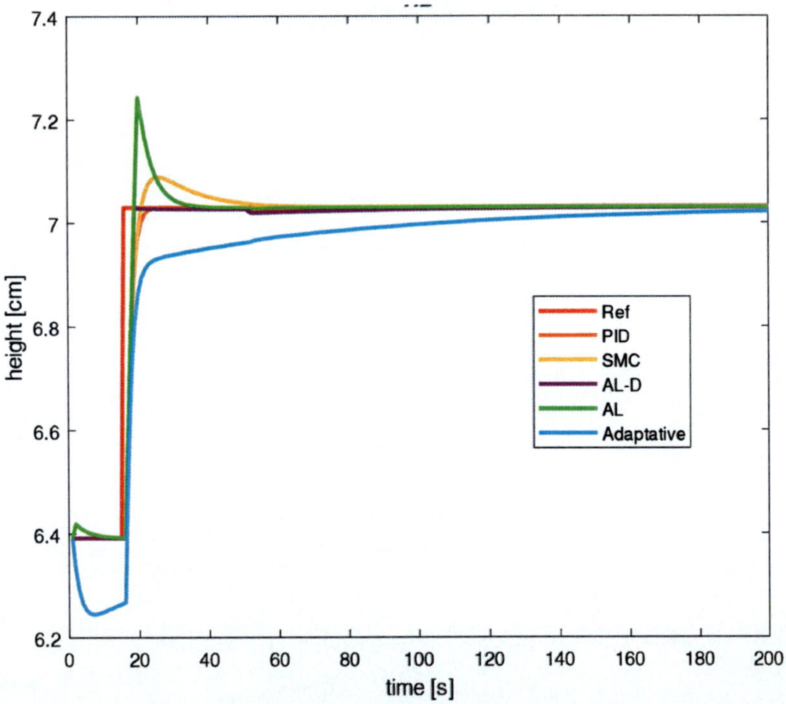

Fig. 5. Step response H2

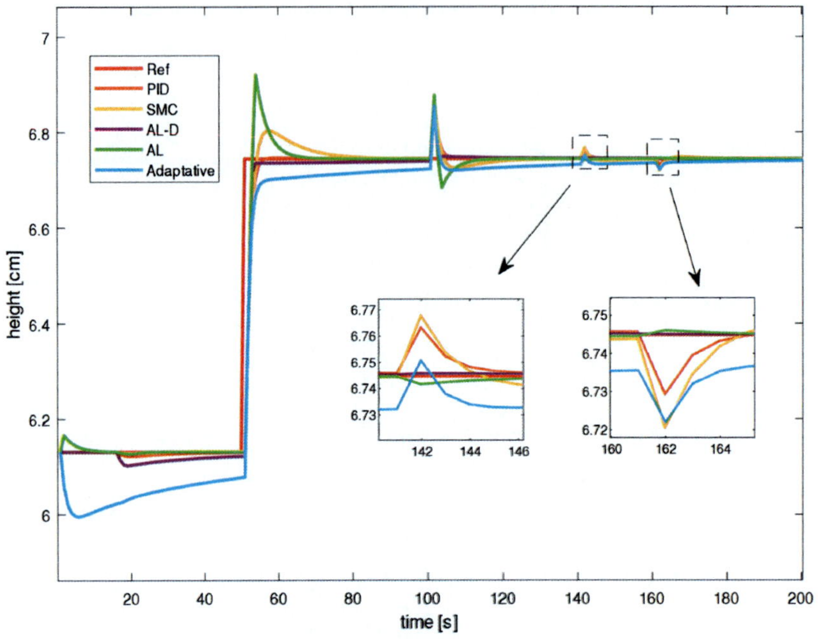

Fig. 6. Disturbance is rejection H1

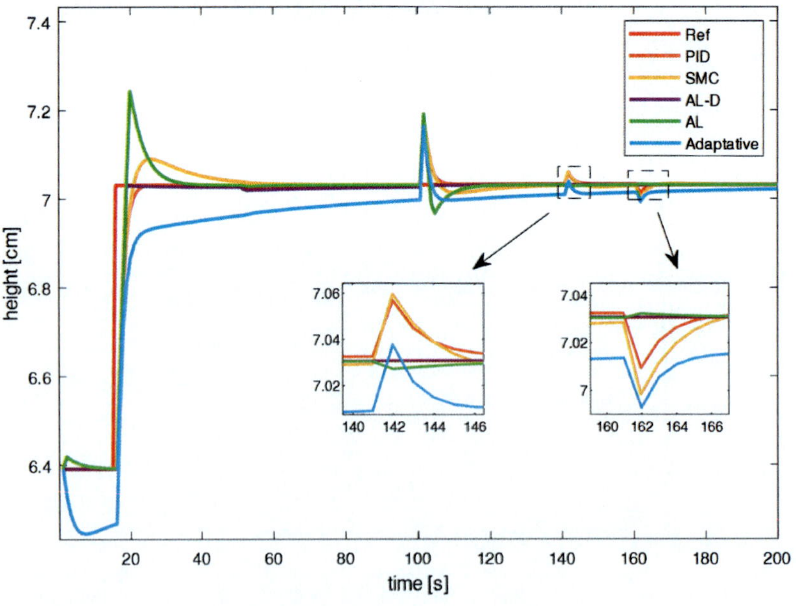

Fig. 7. Disturbance is rejection H2

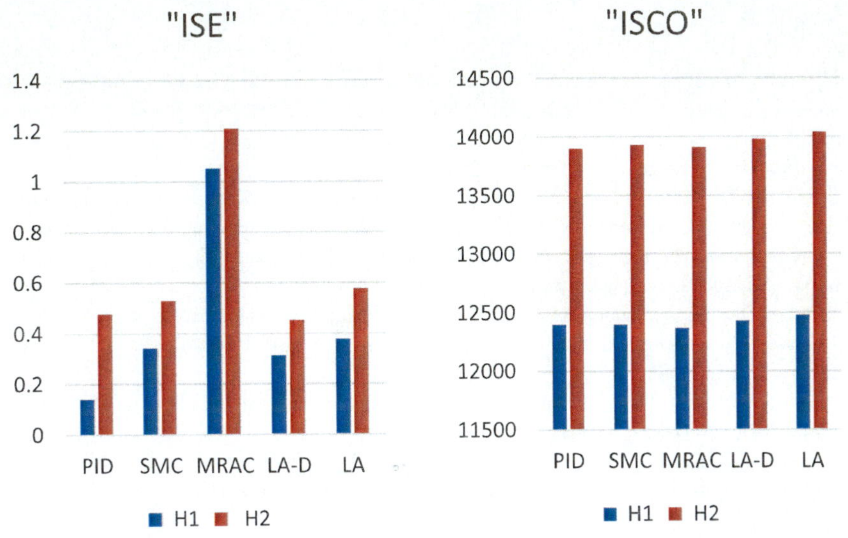

Fig. 8. Performance indices

Applying disturbances of 30%, 25%, −25% from its final value, it is seen on H1 and H2.

The graphs below show the comparison of 4 tuning methods that had not been performed before, in which the advantage of using linear algebra with and without decouplers is shown, as to the error rate when using decouplers for tank heights 1 and 2.

5 Conclusion

When performing a multivariable control, it is necessary to see the interaction of the output variables with the input variables to determine the models for the design and decide whether or not to perform decouplers for a more inherent control. Three advanced control techniques were successfully compared, which were Sliding Mode Control, Linear Algebra, and Model Reference Adaptative Control, against a PID scheme tuned by Dahlin and implemented with decouplers following the Bristol method. From the results obtained of ISCO and ISE, it can be determined that, for the quadruple tank model, the PID obtains good results for the tests carried out in the event of reference changes and disturbance, obtaining a similar performance to the advanced control techniques with which it was compared. For this reason, for the quadruple tank process, the use of PID schemes with decouplers is still a valid option. For future work, it is suggested to perform the same comparison of control techniques on models that present greater challenges for the PID scheme, such as high nonlinearities, to highlight the qualities of the advanced control schemes like Sliding Mode Control, Linear Algebra, and Model Reference Adaptative Control.

References

1. Camacho, O., Rojas, R., Garcia, W.: Variable Structure Control applied to chemical processes with inverse response. ISA Trans. **38**(1), 55–72 (1999). ISSN 0019-0578
2. Scaglia, G., Serrano, M.E., Albertos, P.: Linear Algebra Based Controllers: Design and Applications. Springer, Heidelberg (2020)
3. Sardella, M.F., Serrano, M.E., Camacho, O., Scaglia, G.J.E.: Design and application of a linear algebra based controller from a reduced-order model for regulation and tracking of chemical processes under uncertainties. Industr. Eng. Chem. Res. **58**, 15222–15231 (2019). https://doi.org/10.1021/acs.iecr.9b01257
4. Scaglia, G., et al.: Linear algebra based controller design applied to a bench-scale oenological alcoholic fermentation. Control Eng. Pract. **25**, 66–74 (2014). https://doi.org/10.1016/j.conengprac.2014.01.002
5. Camacho, O., Rosales, A., Rivas: Control de procesos. Chap. Sintonizac, pp. 125–127. EPN Editorial, Quito (2020). _rst edn
6. Chirag: Simple Adaptive Control Example (2021). https://www.mathworks.com/matlabcentral/fileexchange/44416-simple-adaptive-control-example. MATLAB Central File Exchange. Accessed 6 Dec 2021
7. Serrano, M., Scaglia, G., Abillay, P., Ortiz, O.: Linear algebra and optimization based controller design for trajectory tracking of typical chemical process. Latin Am. Appl. Res. **44**, 313–318 (2014)
8. Johansson, K.H.: The quadruple-tank process: a multivariable laboratory process with an adjustable zero. IEEE Trans. Control Syst. Technol. **8**(3), 456–465 (2000)
9. Raj, R.A., Deepa, S.: Modeling and implementation of various controllers used for Quadruple-Tank. In: 2016 International Conference on Circuit, Power and Computing Technologies (ICCPCT), pp. 1–5. IEEE (2016)
10. Christie, X.P.P., Jose, P.S.: Design of model reference adaptive control for a quadruple tank system using LabVIEW. In: 2015 International Conference on Innovations in Information, Embedded and Communication Systems (ICIIECS) (2015)
11. Herrera, M., Gonzales, O., Leica, P., Camacho, O.: Robust controller based on an optimal - integral surface for quadruple-tank process. In: 2018 IEEE Third Ecuador Technical Chapters Meeting (ETCM) (2018)
12. Ulloa, F., Camacho, O., Leica, P.: A proposal for teaching advanced control techniques using a virtual processes laboratory. In: 2018 IEEE Third Ecuador Technical Chapters Meeting (ETCM), pp. 1–6 (2018). https://doi.org/10.1109/ETCM.2018.8580335

Learning an Improved LMI Controller Based on Takagi-Sugeno Models via Value Iteration

Henry Díaz(✉), Karla Negrete, and Jenyffer Yépez

Universidad Técnica del Norte, Ibarra, Ecuador
{hpdiaz,kpnegrete,jayepez}@utn.edu.ec

Abstract. This article proposes an alternative for formulating the method to improve the conservative controllers based on linear matrix inequality (LMI), action-value function (Q-function), and value iteration algorithm to learn optimal controllers by using system data. In this respect, the proposed uses ideas of the previous works that parametrize in a particular way the Q-function. In this sense, the Q-function can be described with polynomials membership functions for fuzzy models of Takagi-Sugeno and initialize a learning process with the LMI controller. The obtained controller uses both the information about the membership functions and a set of data obtained from the system to improve the LMI controller. A TORA system is used to illustrate the approach.

Keywords: Reinforcement learning · Q function · Approximate dynamic programming · Takagi-Sugeno fuzzy models · Linear matrix inequalities

1 Introduction

Several nonlinear systems could be modeled using Takagi-Sugeno (TS) models by sector nonlinearity modeling technique [16]. A feature of TS models is to express nonlinearities as a convex combination of linear systems. TS model uses the linear matrix inequality (LMI) techniques for designing controllers. These controllers guarantee that the closed-loop fuzzy system has global asymptotic stability [20].

On the other hand, several studies have been published related to reinforcement learning [2,15], neurodynamic programming [1], approximate dynamic programming [14], Q-learning [19], etc. The book by Busoniu [4] presents an accessible presentation of the dynamic programming (DP) and reinforcement learning (RL) methods of finding value function approximators. RL methods have been mathematically expressed as a control perspective with the following practically feasible approaches e.g. [9,11–13].

Combining both nonlinear systems representation using TS form and reinforcement learning approaches is proposed in [7,8]. These works use a nonlinear

M. Botto-Tobar et al. (Eds.): *Latest Advances in Electrical Engineering, and Electronics*, LNEE 933, pp. 89–99, 2022.
https://doi.org/10.1007/978-3-031-08942-8_7

system in TS form and the parallel distributed compensation (PDC) structures associated with them in order to fix a particular parametrization of Q-function. Using the policy iteration (PI) algorithm achieves an improvement in the performance of the LMI controller. Note that the LMI solution is used as initialization of the Q-function, and the Q-function is parameterized with TS membership functions.

The objective of this paper is to increase the applicability of the ideas presented in previous works that parametrize the Q-function with TS membership functions. These works use the PI algorithm to improve the performance of the LMI controller. We propose to use the ideas of parameterization and initialization with another popular reinforcement learning algorithm that corresponds to value iteration (VI) algorithm. In this way, either algorithm can be used interchangeably to improve the performance of the LMI solution.

2 Methodology

2.1 Takagi-Sugeno Systems

The nonlinear dynamic system can be expressed (at least locally) following the sector nonlinearity idea based on TS fuzzy models [16]:

$$x_{t+1} = \sum_{i=1}^{\rho} \mu_i(x_t)(A_i x_t + B_i u_t) \tag{1}$$

with $x_t \in \mathbb{X} \subset \mathbb{R}^{n_x}$ represents state vector and $u_t \in \mathbb{U} \subset \mathbb{R}^{n_u}$ is the control input. In Eq. (1), $\mu(x_t)$ is a set of $\rho = 2^p$ nonlinear membership functions (p denotes de number of nonlinearities). In TS modeling the membership functions fulfills the convex sum condition:

$$\sum_{i=1}^{\rho} \mu_i(x_t) = 1 \tag{2}$$

with $0 \leq \mu_i(x_t) \leq 1$.

The TS model (1) is a representation of vertices of a convex linear problem and the PDC control design methodology can be used. In this context, a value function overbound can be calculated via LMIs:

$$V^{\pi}_{LMI}(x_t) := x_t^T P_{LMI} x_t \geq V^{\pi^*}(x_t). \tag{3}$$

The LMI policy is established as:

$$\pi_{LMI}(x_t) := -\sum_{i=1}^{\rho} \mu_i(x_t) K_i x_t, \tag{4}$$

with $K_i = F_i P_{LMI}^{-1}$, further details of the method can be found in [17].

2.2 Reinforcement Learning and Optimal Control

On the other hand, in the case of arbitrary state feedback policy $u_t := \pi(x_t)$ the value function (V-function) of initial state x_0 is defined as:

$$V^{\pi}(x_0) := \sum_{t=0}^{\infty} \gamma^t r(x_t, u_t) = \sum_{t=0}^{\infty} \gamma^t \left(x_t^T H_x x_t + u_t^T H_u u_t\right) \tag{5}$$

where $\gamma \geq 1$ is a decay rate, r_t is the immediate cost. Weighting matrices $H_x \in \mathbb{R}^{n_x \times n_x}$ and $H_u \in \mathbb{R}^{n_u \times n_u}$ are related with quadratic cost function.

The design controller $u_t = \pi(x_t)$ should minimized $V^{\pi}(x_t)$ (LQR in the linear case). By considering the action-value function (Q-function) with a fixed control policy $\pi(x_t)$ applying the equivalence $V^{\pi}(x_t) = Q^{\pi}(x_t, \pi(x_t))$ the Q-function can be described in accordance with Bellman's equation as follows:

$$Q^{\pi}(x_t, u_t) = r(x_t, u_t) + \gamma Q^{\pi}(x_{t+1}, \pi(x_{t+1})) \tag{6}$$

Bellman's principle of optimality can be used to derivate the optimal *action-value* function $Q^{\pi^*}(x_t, u_t)$ and its associated control policy:

$$Q^{\pi^*}(x_t, u_t) = r(x_t, u_t) + \gamma \min_{u'} Q^{\pi^*}(x_{t+1}, u') \tag{7}$$

$$u_t^* = \pi^*(x_t) = \arg\min_{u_t} Q^{\pi^*} \tag{8}$$

The two most popular methods to calculate the optimal value function are value and policy iteration algorithms [12]. In the case of VI algorithm computes the optimal value function, and this value function is used to calculate an optimal policy. On the other hand, PI algorithm evaluates policies by building their value functions and computes the improved policies using the previously computed value functions [4,5].

2.3 Reinforcement Learning for Takagi-Sugeno Systems

Previous works [6,8] propose a data-driven algorithm that link the guaranteed-cost solutions with LMIs and the RL schemes in order to learn optimal controllers. The main idea is to initialize the PI algorithm with the LMI solution. We give a brief overview on the proposed method.

In the case of linear discrete-time system $x_{t+1} = Ax_t + Bu_t$ with a state feedback $u_t = \pi(x_t) = -K^{\pi} x_t$.

The quadratic Q-function is:

$$Q^{\pi}(x_t, u_t) := \begin{bmatrix} x_t \\ u_t \end{bmatrix}^T \begin{bmatrix} S_{xx}^{\pi} & S_{xu}^{\pi} \\ S_{ux}^{\pi} & S_{uu}^{\pi} \end{bmatrix} \begin{bmatrix} x_t \\ u_t \end{bmatrix} \tag{9}$$

By applying $\frac{\partial Q^{\pi}(x_t, u_t)}{\partial u_t} = 0$ to (9) we get:

$$\hat{u}_t = -\left(S_{uu}^{\pi}\right)^{-1} S_{ux}^{\pi} x_t \tag{10}$$

Note that $K^{\pi} = \left(S^{\pi}_{uu}\right)^{-1} S^{\pi}_{ux}$ would be the same value as the optimal LQR gain [10].

Considering the polynomials (degree q) of the TS modeling nolinearities the proposed parameterization of the Q-function is as follows:

$$Q^{\pi}_{\mu^q}(x_t, u_t) := \begin{bmatrix} x_t \\ u_t \end{bmatrix}^T S^{\pi}_{\mu^q}(x_t) \begin{bmatrix} x_t \\ u_t \end{bmatrix} \tag{11}$$

with

$$S^{\pi}_{\mu^q}(x_t) : = \begin{bmatrix} S^{\pi}_{\mu^q,xx} & S^{\pi}_{\mu^q,xu} \\ S^{\pi}_{\mu^q,ux} & S^{\pi}_{\mu^q,uu} \end{bmatrix} \tag{12}$$

and the policy:

$$\hat{\pi}_{\mu^q}(x_t) := -(S^{\pi}_{\mu^q,uu}(x_t))^{-1} S^{\pi}_{\mu^q,ux}(x_t) x_t, \tag{13}$$

being a state feedback control policy that relies on membership functions.

The Q-function parametrization using a polynomial degree-2 in equation (11) and the LMI solution is expressed as the following formula:

$$Q^{\pi}_{\mu^2}(x_t, u_t) = \begin{bmatrix} x_t \\ u_t \end{bmatrix}^T S^{\pi}_{\mu^2}(x_t) \begin{bmatrix} x_t \\ u_t \end{bmatrix} \tag{14}$$

with,

$$S^{\pi}_{\mu^2}(x_t) := \begin{bmatrix} S^{\pi}_{\mu^2,xx}(x_t) & S^{\pi}_{\mu^2,xu}(x_t) \\ S^{\pi}_{\mu^2,ux}(x_t) & S^{\pi}_{\mu^2,uu}(x_t) \end{bmatrix} \tag{15}$$

$$S^{\pi}_{\mu^2,xx}(x_t) := H_x + \gamma \sum_{i=1}^{\rho} \mu_i(x_t) A_i^T \cdot P_{LMI} \cdot \sum_{i=1}^{\rho} \mu_i(x_t) A_i \tag{16}$$

$$S^{\pi}_{\mu^2,xu}(x_t) := \gamma \sum_{i=0}^{\rho} \mu_i(x_t) A_i^T \cdot P_{LMI} \cdot \sum_{i=1}^{\rho} \mu_i(x_t) B_i \tag{17}$$

$$S^{\pi}_{\mu^2,uu}(x_t) := \gamma \sum_{i=1}^{\rho} \mu_i(x_t) B_i^T \cdot P_{LMI} \cdot \sum_{i=1}^{\rho} \mu_i(x_t) B_i + H_u \tag{18}$$

where the matrix P_{LMI} overbounds the optimal value function, so that, the Q-function above denotes an upper bound of the true Q-function.

The resulting controller is described as the one-step controller and can be described as:

$$\hat{\pi}_1(x_t) := -(S^{\pi}_{\mu^2,uu}(x_t))^{-1} (S^{\pi}_{\mu^2,xu}(x_t))^T x_t. \tag{19}$$

In this context, the PI algorithm can be applied with this specific parametrization of Q-function. In addition, most RL frameworks suggest to learn a specific Q-function parameterization based on regressors (i.e. linear parametrization).

Using a linear parametrization:

$$Q^{\pi}(x_t, u_t) \approx (\omega^{\pi})^T \varphi(x_t, u_t) \tag{20}$$

with $\varphi(x_t, u_t) : \mathbb{X} \times \mathbb{U} \mapsto \mathbb{R}^m$. Being a set of m regressors and $\omega^{\pi} \in \mathbb{R}^{m \times 1}$ the weights to be learned.

Considering any admissible (i.e. stabilizing) control policy $\pi(x_t)$ and a Q-function computed with (20). In addition, N samples of an experiment are gathered. The weights in $Q^{\pi}(x_t, u_t)$ can be learned from the data and the PI algorithm. PI algorithm is an interleave firstly policy evaluation and secondly policy improvement.

The Eq. (6) using the linear parametrization is:

$$(\omega^{\pi})^T(\varphi(x_t, u_t)) = r(x_t, u_t) + (\omega^{\pi})^T \gamma(\varphi(\bar{x}_{t+1}, \pi(\bar{x}_{t+1}))) \tag{21}$$

Policy Evaluation. Using the least-squares fitting approximation of $Q^{\pi}(x_t, u_t)$ is obtained of (21) by setting

$$\omega^{\pi} = \left[\Phi - \gamma \Phi^{\pi +}\right]^{\dagger} \cdot R, \tag{22}$$

where $\dagger$ refers the Moore-Penrose pseudoinverse.

Policy Improvement. The improved policy step can be computed by solving:

$$\hat{u}_t := \hat{\pi}(x_t) := \arg\min_{u_t} Q^{\pi}(x_t, u_t), \tag{23}$$

The main result is a shape-dependent controller that improves the LMI solution. In this sense, the LMI solution provides a good initialization of the approximated Q-function and the PI algorithm improves the controller. Additionally, the work of [7] proposed a modification to previous Q-function parametrization by introducing the gradient estimation of the nonlinearity. As result, the learning controller shows improved performance over LMI controller and the controller obtained with the Q-function parametrization without the gradient information.

Considering the previous ideas, the objective of the paper is to extend the applicability of the proposed parameterization of the Q-function initialized with the LMI solution to another of the most used algorithms in reinforcement learning, which corresponds to the VI algorithm.

2.4 Value Iteration for Takagi-Sugeno Systems

VI algorithm uses the Bellman optimality equation to evaluate an optimal value funtion. The optimal policy is derived of the optimal value function. The VI algorithm can be used considering the Q-function with approximation of (20) and parametrization of (11) with polynomial degree-2.

Value Update. Update the Q-function using

$$Q^{\pi}{}_{j+1}(x_t, u_t) = r(x_t, u_t) + \gamma Q^{\pi}{}_j(x_{t+1}, \pi(x_{t+1})). \tag{24}$$

$$(\omega^{\pi}_{j+1})^T(\varphi(x_t, u_t)) = r(x_t, u_t) + (\omega^{\pi}_j)^T \gamma(\varphi(\bar{x}_{t+1}, \pi(\bar{x}_{t+1}))) \tag{25}$$

Using the least-squares approximation of $Q^{\pi}(x_t, u_t)$ is obtained of (21) by setting

$$\omega^{\pi}_{j+1} = [\Phi]^{\dagger} \cdot [R + \gamma\omega^{\pi}_j \Phi^{\pi+}] \tag{26}$$

Policy Improvement. An improved policy can be determined by using (23).

The learning algorithm summarize the procedure:

Learning Algorithm

1. Generate a dataset with N random samples $(x_t,\ u_t)$. The dataset can be obtained by simulation or experimental.
2. Get a TS model
3. Define the membership functions associated to TS model.
4. Determine the cost function.
5. Compute an LMI upper cost bound.
6. Initialized a controller using the LMI solution.
7. Update Q-function using (25)
8. Improve the policy with (26).
9. Iterate step 6 and 7 until convergence (with the same dataset).

3 Results and Discussion

3.1 Numerical Example

The proposed methodology is applied on a TORA system (Translational Oscillations control with a Rotational proof mass Actuator). A representation of the system is shown in Fig. 1. Note that TORA system is connected by a linear spring to fix a wall. The cart is constrained to movement in one dimension.

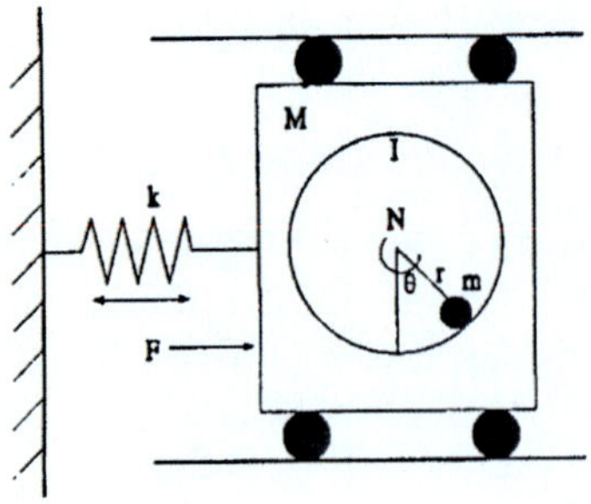

Fig. 1. A schematic representation of TORA system

Let x_1 denotes the cart translational position and the cart velocity is $x_2 = \dot{x_1}$. Let $x_3 = \theta$ and $x_4 = \dot{x_3}$ indicate the angular position and velocity of the rotational proof mass. The system dynamics can be described by the equation (27) [3]. The torque applied on the eccentric mass is u. Note that we assume there is no disturbance.

$$\dot{x} = f(x) + g(x)u \tag{27}$$

with

$$f(x) = \begin{bmatrix} x_2 \\ \frac{-x_1+\epsilon x_4^2(\sin x_3)}{1-\epsilon^2(\cos^2 x_3)} \\ x_4 \\ \frac{\epsilon(\cos x_3)(x_1+\epsilon x_4^2 \sin x_3)}{1-\epsilon^2(\cos^2 x_3)} \end{bmatrix} \tag{28}$$

$$g(x) = \begin{bmatrix} 0 \\ \frac{\epsilon cos x_3}{1-\epsilon^2(\cos^2 x_3)} \\ 0 \\ \frac{1}{1-\epsilon^2(\cos^2 x_3)} \end{bmatrix} \tag{29}$$

Using state transformation and the feedback transformation [18]:

$$\begin{aligned} z_1 &= x_1 + \epsilon \sin x_3 \\ z_2 &= x_2 + \epsilon x_4 \cos x_3 \\ y_1 &= x_3 \\ y_2 &= x_4 \end{aligned}$$

$$\begin{aligned} v = \frac{1}{1-\epsilon^2 \cos^2 y_1}&[\epsilon \cos y_1(z_1 - (1 + y_2^2)\epsilon \sin y_1) + u] \\ &= \alpha(z_1, y_2 + \beta(y_1)u) \end{aligned}$$

We have the following form:

$$\begin{aligned} \dot{z}_1 &= z_2 \\ \dot{z}_2 &= -z_1 + \epsilon \sin y_1 \\ \dot{y}_1 &= y_2 \\ \dot{y}_2 &= v \end{aligned}$$

Considering that the equilibrium point is any point in $[0, 0, x_3, 0]$ but the desired equilibrium point is $[0, 0, 0, 0]$. The TORA system representation considering one nolinearity is:

$$\begin{bmatrix} \dot{z}_1 \\ \dot{z}_2 \\ \dot{y}_1 \\ \dot{y}_2 \end{bmatrix} = \begin{bmatrix} 0 & 1 & 0 & 0 \\ -1 & 0 & \epsilon\frac{\sin(y_1)}{y_1} & 0 \\ 0 & 0 & 0 & 1 \\ 0 & 0 & 0 & 0 \end{bmatrix} \begin{bmatrix} z_1 \\ z_2 \\ y_1 \\ y_2 \end{bmatrix} + \begin{bmatrix} 0 \\ 0 \\ 0 \\ 1 \end{bmatrix} v \tag{30}$$

The nonlinearity in Eq. (30) is given by the sinusoidal function. For the TS model, the values of membership functions are computed by:

$$\mu_1(y) := \begin{cases} \frac{\sin y/y - \sin(y_{\max})/y_{\max}}{1-\sin(y_{\max})/y_{\max}} & \text{if } y \neq 0 \\ 1 & \text{if } y = 0 \end{cases} \tag{31}$$

$$\mu_2(y) := 1 - \mu_1(y) \tag{32}$$

with $y_t \in [-y_{\max}, y_{\max}]$ rad. and $y_{\max} = \pi$. Model vertex matrices discretized with forward Euler approximation are:

$$A_1 = \begin{bmatrix} 1 & \delta & 0 & 0 \\ -\delta & 1 & 0 & 0 \\ 0 & 0 & 1 & \delta \\ 0 & 0 & 0 & 1 \end{bmatrix}, \qquad B_1 = \begin{bmatrix} 0 \\ 0 \\ 0 \\ \delta \end{bmatrix},$$

$$A_2 = \begin{bmatrix} 1 & \delta & 0 & 0 \\ -\delta & 1 & \delta\epsilon & 0 \\ 0 & 0 & 1 & \delta \\ 0 & 0 & 0 & 1 \end{bmatrix}, \qquad B_2 = \begin{bmatrix} 0 \\ 0 \\ 0 \\ \delta \end{bmatrix}.$$

with the sampling time $\delta = 0.01$ and $\epsilon = 0.1$. The range of the state space is $z_1 \in [-1, 1]$, $z_2 \in [-1, 1]$, $y_1 \in [-\pi, \pi]$, $y_2 \in [-4, 4]$ and the control signal $v \in [-50, 50]$.

For $H_x = \text{diag}([100, 1, 10, 1])$, $H_u = 1$, $\gamma = 1$ provides the following policy (LMI controller):

$$\hat{\pi}_{LMI}(x_t) = -\,[97.8281 \;\; -228.5067 \;\; -16.1794 \;\; -6.0419]\, x_t. \tag{33}$$

The LMI controller provides the following bound for the value function:

$$\bar{V}^{\pi_{LMI}}(x) = x^T \cdot 10^6 \begin{bmatrix} 3.1439 & -0.4655 & -0.0322 & -0.0101 \\ -0.4655 & 4.7546 & 0.1176 & 0.0241 \\ -0.0322 & 0.1176 & 0.0078 & 0.0017 \\ -0.0101 & 0.0241 & 0.0017 & 0.0006 \end{bmatrix} \cdot x \tag{34}$$

The initialization of the proposed learning algorithm is the LMI controller and the dataset $\mathcal{D}$ with $N = 29 \times 10^3$ data points. As result, the second-order rational controller is:

$$\hat{\pi}_{\mu^2}(x_t) = -10^3 \begin{bmatrix} \frac{-0.9348\mu_1^2(x_t) - 0.4149\mu_1(x_t) + 2.2351}{0.4237\mu_1^2(x_t) - 0.1778\mu_1(x_t) - 1.3064} \\ \frac{0.9043\mu_1^2(x_t) - 0.1702\mu_1(x_t) - 4.1601}{0.4237\mu_1^2(x_t) - 0.1778\mu_1(x_t) - 1.3064} \end{bmatrix}^T \cdot x_t. \tag{35}$$

Figure 2 shows trajectories using both LMI controller and RL proposed controller. As seen in this figure the learned controller is stabilizing. Figure 3 represents the accumulated cost that in this example the learned controller cost (330) is less that LMI controller cost (1100).

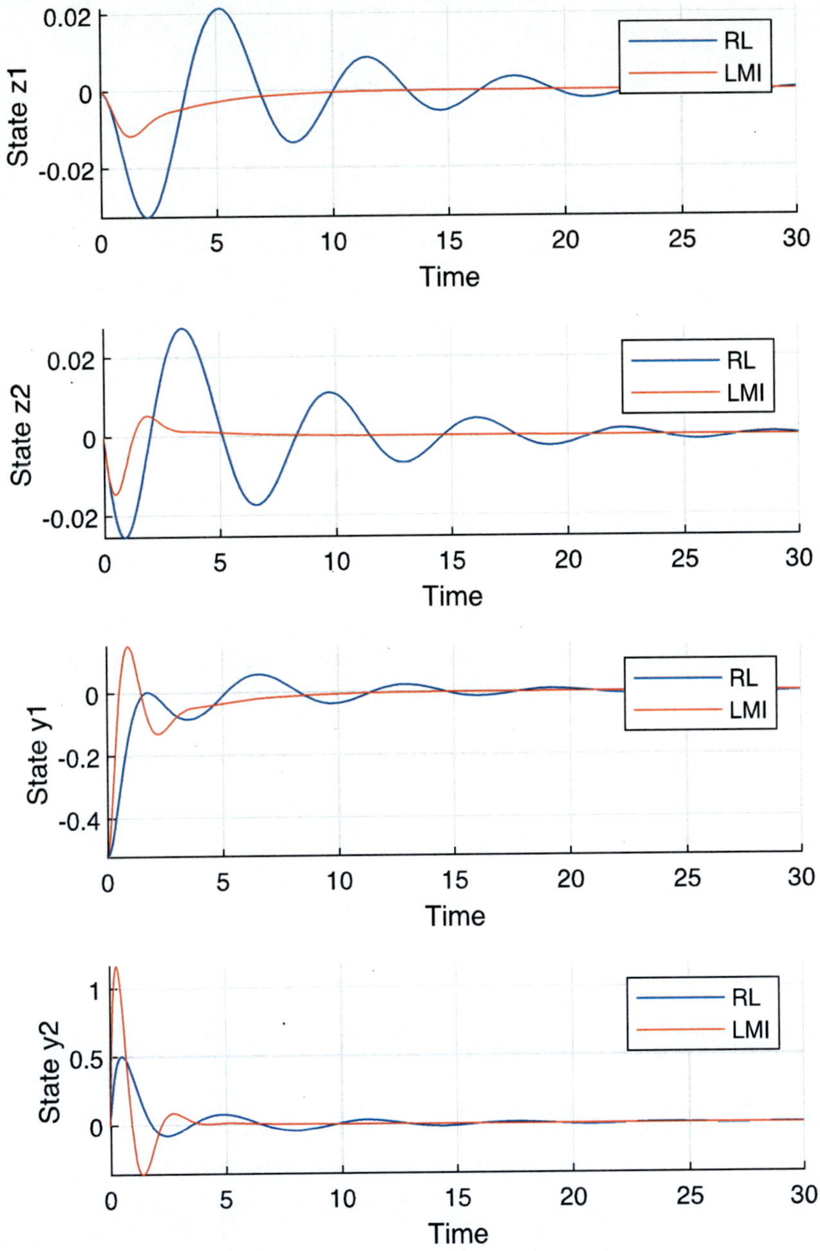

Fig. 2. Controller trajectory from the initial state $x_0 = [0\ 0\ 0.53\ 0]^T$. Learned controller in blue and the LMI controller in red. (Color figure online)

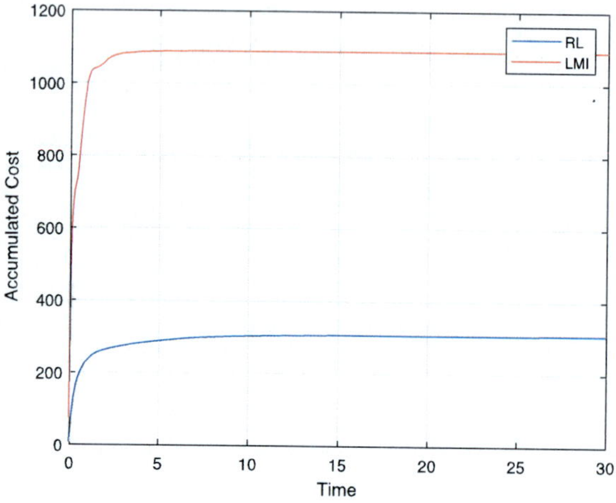

Fig. 3. Accumulated cost of learned controller (blue) and the LMI controller (red). (Color figure online)

4 Conclusions

This paper uses the ideas that link the conservative results from guaranteed-cost LMIs with RL approach. In previous work, a parameterization of the action-value function (Q-function) is implemented using TS membership functions and the PI algorithm is initialized with the LMI controller. This proposed achieves to improve the performance of the LMI controller. The same approach has been implemented using another popular RL algorithm which is the VI algorithm. The proposed algorithm with this parameterization and initialization has converged to a stabilizing controller using a dataset obtained by simulation of the TORA nonlinear system. In this way, both the PI and VI algorithms can be used interchangeably to improve the performance of the LMI solution.

References

1. Bertsekas, D.P., Tsitsiklis, J.N.: Neuro-Dynamic Programming. Athena Scientific, Belmont (1996)
2. Bertsekas, D.P.: Reinforcement Learning and Optimal Control. Athena Scientific, Belmont (2019)
3. Bupp, R.T., Bernstein, D.S., Coppola, V.T.: A benchmark problem for nonlinear control design (1998)
4. Busoniu, L., Babuska, R., De Schutter, B., Ernst, D.: Reinforcement Learning and Dynamic Programming Using Function Approximators. CRC Press, Boca Raton (2010)
5. Busoniu, L., de Bruin, T., Tolic, D., Kober, J., Palunko, I.: Reinforcement learning for control: Performance, stability, and deep approximators. Ann. Rev. Control **46**, 8 – 28 (2018). https://doi.org/10.1016/j.arcontrol.2018.09.005, http://www.sciencedirect.com/science/article/pii/S1367578818301184

6. Díaz, H., Armesto, L., Sala, A.: Fitted q-function control methodology based on Takagi-Sugeno systems. IEEE Trans. Control Syst. Technol. **28**, 1–12 (2018). https://doi.org/10.1109/TCST.2018.2885689
7. Díaz, H., Sala, A., Armesto, L.: Improving LMI controllers for discrete nonlinear systems using policy iteration. In: 2017 21st International Conference on System Theory, Control and Computing (ICSTCC), pp. 833–838 (2017). https://doi.org/10.1109/ICSTCC.2017.8107140
8. Díaz, H., Armesto, L., Sala, A.: Improvement of LMI controllers of takagi-sugeno models via q-learning. IFAC PapersOnLine **49**(5), 67–72 (2016). https://doi.org/10.1016/j.ifacol.2016.07.091, http://www.sciencedirect.com/science/article/pii/S2405896316302877
9. Kiumarsi, B., Lewis, F.L., Modares, H., Karimpour, A., Naghibi-Sistani, M.B.: Reinforcement q-learning for optimal tracking control of linear discrete-time systems with unknown dynamics. Automatica **50**(4), 1167–1175 (2014)
10. Lewis, F., Vrabie, D., Syrmos, V.: Optimal Control, 3rd edn. John Wiley & Sons, Hoboken (2012)
11. Lewis, F.L., Liu, D.: Reinforcement Learning and Approximate Dynamic Programming for Feedback Control. Wiley, Hoboken (2013)
12. Lewis, F.L., Vrabie, D.: Reinforcement learning and adaptive dynamic programming for feedback control. IEEE Circ. Syst. Maga. **9**(3), 32–50 (2009)
13. Lewis, F.L., Vrabie, D., Vamvoudakis, K.G.: Reinforcement learning and feedback control: using natural decision methods to design optimal adaptive controllers. IEEE Control Syst. **32**(6), 76–105 (2012)
14. Powell, W.B.: Approximate Dynamic Programming: Solving the Curses of Dimensionality. Wiley, Hoboken (2011)
15. Sutton, R.S., Barto, A.G.: Reinforcement Learning: An Introduction, 2nd edn. MIT press, Cambridge (2018)
16. Tanaka, K., Wang, H.O.: Fuzzy Control Systems Design and Analysis: A Linear Matrix Inequality Approach. John Wiley & Sons, Hoboken (2001)
17. Tuan, H.D., Apkarian, P., Narikiyo, T., Yamamoto, Y.: Parameterized linear matrix inequality techniques in fuzzy control system design. IEEE Trans. Fuzzy Syst. **9**(2), 324–332 (2001)
18. Wang, H., Li, J., Durham, N.: Nonlinear control via PDC: the TORA system example (2007)
19. Watkins, C.J., Dayan, P.: Q-learning. Mach. Learn. **8**(3–4), 279–292 (1992)
20. Wu, H.N., Cai, K.Y.: H2 guaranteed cost fuzzy control for uncertain nonlinear systems via linear matrix inequalities. Fuzzy Sets Syst. **148**(3), 411–429 (2004)

A Sliding Mode Controller Approach Based on Particle Swarm Optimization for ECG and Heart Rate Tracking Purposes

Sebastián Estrada[1], Jorge Espín[1], Leandro Ponce[1(✉)], Mishell Espín[2], and Jorge Estrada[3]

[1] Escuela Politécnica Nacional, Quito, Ecuador
{juan.estrada01,jorge.espin01,leandro.ponce}@epn.edu.ec
[2] Universidad Central del Ecuador, Quito, Ecuador
bmespin@uce.edu.ec
[3] Hospital de Especialidades "Eugenio Espejo", Quito, Ecuador
j.estrada@heee.gob.ec
https://www.epn.edu.ec, https://www.uce.edu.ec/web/fcm, http://hee.gob.ec

Abstract. This paper aims to improve the design of two Sliding Mode Controllers (SMC); one of them based on a first-order model and the other one based on an integrating model. The proposed controllers are tuned using the PSO (Particle Swarm Optimization) algorithm in order to minimize performance indexes and improve transient characteristics. To show these are promising alternatives, they are compared to traditional PID controllers. The suggested controllers are applied for tracking purposes in cardiac models, yielding a satisfactory performance over an ECG and a variable heart rate reference, proven by lower ISE (Integral of Squared Error) and TVu (Total Variation of control effort) values.

Keywords: SMC · Heart models · Pso algorithm · ECG tracking · Integrating systems

1 Introduction

Cardiac pacemakers are being used more and more frequently. It is estimated that more than 3.5 million have been implanted worldwide and approximately 700,000 devices are implanted annually [5]. For this reason, research and continuous improvement of these devices have become a focal point for the research community.

The cardiac pacemaker is an energy supplier that controls the duration, intensity, and rhythm in order to stimulate the heart appropriately. It is responsible for maintaining an ideal ventricular rate, especially in bradycardia disorders. Electrophysiologically, it contributes to the treatment of certain tachycardia, in addition to restoring and increasing the cardiac frequency according to the needs

M. Botto-Tobar et al. (Eds.): *Latest Advances in Electrical Engineering, and Electronics*, LNEE 933, pp. 100–114, 2022.
https://doi.org/10.1007/978-3-031-08942-8_8

of the human body [6]. For patients with heart rhythm disorders, implantation of a cardiac pacemaker is recommended.

Heart movement starts when the myocardium generates electrical impulses that stimulate the contraction of the muscle. This electrical signal originates in the Sinoatrial Node (SAN) known as the heart's "natural pacemaker". The electrical impulse contracts the atria and the signal passes through the atrioventricular node, which stops the signal for a brief moment and sends it through the muscle fibers of the ventricles, stimulating their contraction. The SAN sends electrical impulses with a specific frequency; however, it can vary according to physical demands, stress levels, hormonal or pathological factors. Malfunctioning of the SAN can cause the heart to beat too fast, too slow, or irregularly. In some circumstances, the conduction pathways are blocked, resulting in an irregular heart rhythm [11].

In order to control artificial heart models, control techniques such as SMC and PID can be applied to perform ECG tracking and heart rate stabilization. These controllers could be implemented in an artificial pacemaker where heart dynamics are represented using mathematical models such as Zeeman's.

We chose SMC due to its robustness, its low susceptibility to modeling mismatches and its effectiveness in systems with complex dynamics [20]. Since heart models have a high degree of non-linearity, SMC arises as a promising control alternative.

In this paper, an SMC-based control model that simulates cardiac electrical activity is presented, with heart rate and rhythm manipulation capabilities. These control schemes contribute to understanding heart disorders in simulated scenarios, allowing to test and validate better treatments.

Paper outline:

- Section 2 describes SMC and PSO algorithm foundations.
- Section 3 details the heart models used in this work.
- Section 4 focuses on the proposed controllers design.
- Section 5 presents the simulation results.
- Section 6 exposes our conclusions.

2 Background

2.1 Sliding Mode Control

SMC's development started in the 1960s s with the research published on Variable Structure Control (VSC) [19]. VSC aimed to improve the control characteristics on systems that posed a challenge for classical control methods because of uncertainties in the system modeling and disturbances. It consists of a switching method between continuous subsystems resulting in a very robust control structure [10]. From VSC, SMC was derived as an improvement applicable to linear and non-linear systems. SMC considers modeling uncertainties and disturbances as part of the controller design, resulting in a controller best suited for systems where classic controllers are insufficient.

SMC proposes the design of a sliding surface that allows the process state to approach the desired value. The reaching mode takes the system state near the sliding surface. Once the system state has crossed the surface, it switches, and the process enters in a sliding mode where the controlled variable converges to the desired value [3].

The sliding surface design must represent the desired system state behavior. As with any control scheme, the final goal is to stabilize the system state on the desired value. One of the design considerations for the surface S is that the sliding principle, detailed in Eq. (1), must be fulfilled so the surface maintains a constant value.

$$\frac{dS(t)}{dt} = 0 \tag{1}$$

SMC control law consists of a continuous function that depends on the controlled variable and the reference value; and a discontinuous part that includes the switching element of the control law. One of the problems that arise from the use of SMC is control chattering. This issue can be addressed with the tuning parameters of the discontinuous part of the SMC law [17].

2.2 Particle Swarm Optimization Algorithm

The Particle Swarm Optimization (PSO) algorithm was developed by observing and modeling social interactions in the animal kingdom, such as bird flocks and fish schools, where individuals cooperate to achieve a common goal that benefits the group. This model resulted in a useful algorithm where individuals, called particles, represent possible sets of parameters that need to be optimized. The swarm consists of a finite number of particles that move in a multi-dimensional space by considering their own and the group's experience [21]. This experience is translated into the values of the variables to be found in order to optimize an objective function. The swarm is often initialized with a set of random solutions, or particles, which are assigned random velocities.

PSO's success relies on the exchange of information between particles. The information shared between each particle is the individual best experience, or best solution, in regard to the objective function. This information is used in each particle to compare its best result to the group's best solution. In order to search for the best values and optimize the objective function, each particle moves considering three parameters: its personal best solution (pbest), the group's best solution (gbest) and its current velocity (v_i) [13].

On each iteration of the algorithm, each one of the particles moves a certain amount depending on pbest, gbest, and v_i. In general, the algorithm can be described by Eq. (2).

$$(x_i)_{t+1} = (x_i)_t + (v_i)_{t+1} \tag{2}$$

where (x_i) is the position of each particle i. The new velocity of each particle $(v_i)_{t+1}$ is defined in Eq. (3).

$$(v_i)_{t+1} = w * (v_i)_t + c_1 * r_1((pbest_i)_t - (x_i)_t) + c_2 * r_2(gbest_t - (x_i)_t) \tag{3}$$

where the parameters c_1 and c_2 are positive numbers known as cognitive and social weights, respectively. Parameters r_1 and r_2 are random numbers uniformly distributed in the range [0, 1]. Finally, w is known as the inertia weight [2].

3 Heart Models

3.1 Zeeman's Heartbeat Model

The human heartbeat consists of two stages, systole and diastole, where the heart tissues contract and relax respectively. The cycle starts with the sinoatrial node (SAN), located at the top of the right atrium. An electro-chemical signal is generated from the sinoatrial node that propagates through the atrium, causing contraction of the muscle fibers, which in turn, push blood into the lower ventricles. Then, as the electro-chemical signal spreads, the ventricles contract into the systolic state, pushing blood into the arteries. The diastolic state sets once the muscle fibers had relaxed [18].

Heart models aim to represent in a mathematical way these dynamics. Proposed as a second-order non-linear model, a set of equations can describe the three characteristics of the human heartbeat: A stable state (diastole), a threshold to trigger systole, and the return from systole to diastole. Equations (4) and (5) are proposed in [23].

$$\epsilon \dot{x_1} = -(x_1^3 - Tx_1 + x_2), T > 0 \tag{4}$$

$$\dot{x_2} = (x_1 - x_d) - (x_d - x_s)u \tag{5}$$

where ϵ represents a small positive constant associated with the fast eigenvalue of the system. x_1 is associated with the length of the heart's muscle fibers and x_2 is related to chemical control or membrane potential. The value of x_d is a scalar representing the average length of muscle fiber in the diastolic state, and x_s represents a typical fiber length in systolic state. Finally, $T > 0$ expresses tension on the muscle fibers and u is the cardiac pacemaker control signal [14].

3.2 Heart Rate Model

There have been many strategies to model heart dynamics; one of them is the transfer function approach, which aims to represent in a meaningful way the heart-rate stabilization system. The heart-rate model is useful in conjunction with the pacemaker model since both are usually needed to control a patient's heart rate.

To model the pacemaker, a simple low-pass filter transfer function, detailed in Eq. (6), is enough, since the pacemaker's task is to allow some impulses to pass while rejecting high-frequency pulses [22]. For the heart-rate model, the Laplace representation as shown in Eq. (7) can be used.

$$G_p(s) = \frac{8}{s+8} \tag{6}$$

$$\frac{X(s)}{F(s)} = \frac{1}{Ms^2 + Bs + K} \tag{7}$$

where M corresponds to the mass of cardiac muscle, B and K are values proportional to the viscous and torsional drag of myocardial cells, respectively. The force exerted by the heart tissues is represented by F. As a result of the muscle fiber's distortion, a displacement, represented by X is caused. By comparing Eq. (7) with the standard second order model, a final heart-rate model is achieved in Eq. (8) [12].

$$G_h(s) = \frac{w_n^2}{s(s + 2\epsilon w_n)} = \frac{158}{s^2 + 20.11s} \tag{8}$$

A similar model can be used, as shown in [8], yielding a transfer function with similar values. For simulation purposes, Eq. (8) will be used.

4 Controller Design

To control the process outlined above, we propose an SMC based linear model; the parameters will be obtained using the PSO algorithm. Two SMCr were developed, one for first-order systems and the other for integrating systems. The proposed scheme is shown in Fig. 1.

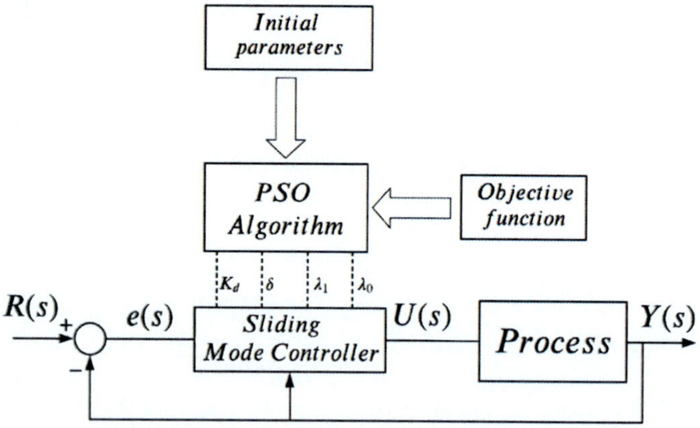

Fig. 1. SMC based on PSO control scheme

4.1 Sliding Mode Controller Based on PSO for First-Order Models

Zeeman's heartbeat model is a highly nonlinear system, given the complex dynamics an SMC controller based on PSO is proposed to regulate the process; the control scheme is shown in Fig. 1. The system's step response can be approximated to the one of a first-order model that has the transfer function shown in Eq. (9).

$$\frac{Y(s)}{U(s)} = \frac{K}{\tau s + 1} \tag{9}$$

In order to obtain the control law, Eq. (9) is represented as a differential equation as shown in Eq. (10):

$$\frac{dY(t)}{dt} = -\frac{1}{\tau}Y(t) + \frac{K}{\tau}U(t) \tag{10}$$

Since the system is modeled as a first-order differential equation, using the error integral as the variable of interest, the sliding surface is defined by:

$$S(t) = e(t) + \lambda \int_0^t e(t)dt \tag{11}$$

The equivalent control law is obtained by solving the sliding condition $\frac{dS(t)}{dt} = 0$, so Eq. (12) is obtained:

$$\frac{dS(t)}{dt} = \frac{de(t)}{dt} + \lambda e(t) \tag{12}$$

Considering that $e(t) = R(t) - Y(t)$, thus its derivative: $\frac{de(t)}{dt} = \frac{dR(t)}{dt} - \frac{dY(t)}{dt}$ and replacing in Eq. (12):

$$\frac{dS(t)}{dt} = \frac{dR(t)}{dt} - \frac{dY(t)}{dt} + \lambda e(t) = 0 \tag{13}$$

Using the model information from Eq. (10), and replacing in Eq. (13):

$$\frac{dS(t)}{dt} = \frac{dR}{dt} - \left[\frac{K}{\tau}U(t) - \frac{1}{\tau}Y(t)\right] + \lambda e(t) \tag{14}$$

Solving Eq. (14) for $U(t)$ the continuous control law is found:

$$U_{eq}(t) = \frac{\tau}{K}\frac{dR(t)}{dt} + \frac{1}{K}Y(t) + \frac{\lambda\tau}{K}e(t) \tag{15}$$

The derivative $\frac{dR(t)}{dt}$ can be eliminated without affecting the controller's performance, thus simplifying the continuous control law:

$$U_{eq}(t) = \frac{1}{K}Y(t) + \frac{\lambda\tau}{K}e(t) \tag{16}$$

Finally, the discontinuous control law is shown in Eq. (17), the sigmoid function is used to overcome the chattering problem.

$$U_d = K_d\frac{S(t)}{|S(t)| + \delta} \tag{17}$$

The final SMC-FO control law is obtained by combining the continuous and discontinuous control laws found in Eq. (16) and in Eq. (17):

$$U(t) = \frac{1}{K}Y(t) + \frac{\lambda\tau}{K}e(t) + K_d\frac{S(t)}{|S(t)| + \delta} \tag{18}$$

SMC-FO has the following tuning parameters: K_d, δ and λ. Criteria to find the optimum parameters is that they should simplify the control law, based on this approach λ is found with:

$$\lambda = \frac{K}{\tau} \tag{19}$$

The other parameters were found using PSO. The objective function is a weighted average value of performance indexes ISE and TVu; giving ISE a higher weight in order to enhance the controller's tracking capability but without disregarding the control effort.

4.2 Sliding Mode Controller Based on PSO for Integrating Systems

The Heart Rate model displays a high-order integrating behavior. Being a complex dynamics system, a Sliding Mode Controller based on the PSO algorithm is proposed as a control alternative.

Sliding Mode Controller for Integrating Systems'(SMC-IS) development is based on a first-order integrating model as shown in Eq. (20).

$$\frac{Y(s)}{U(s)} = \frac{K}{s(\tau s + 1)} \tag{20}$$

Equation (20) can be expressed in differential equation form as shown in Eq. (21).

$$\tau \frac{d^2Y(t)}{dt^2} + \frac{dY(t)}{dt} = KU(t) \tag{21}$$

The sliding surface is defined by:

$$S(t) = \left(\frac{d}{dt} + \lambda\right)^n \int_0^t e(t)\, dt \tag{22}$$

Since Eq. (21) is a second-order differential equation, $n = 2$. Replacing n in Eq. (22) yields Eq. (23).

$$S(t) = \frac{de(t)}{dt} + \lambda_1 e(t) + \lambda_0 \int_0^t e(t)\, dt \tag{23}$$

where $\lambda_1 = 2\lambda$ and $\lambda_0 = \lambda^2$.
In order to fulfill the sliding principle $\frac{dS(t)}{dt} = 0$, the following is obtained:

$$\frac{dS(t)}{dt} = \frac{d^2e(t)}{dt^2} + \lambda_1 \frac{de(t)}{dt} + \lambda_0 e(t) = 0 \tag{24}$$

The error expression $e(t) = R(t) - Y(t)$ is replaced in the equation above.

$$\frac{d^2R(t)}{dt^2} - \frac{d^2Y(t)}{dt^2} + \lambda_1 \left(\frac{dR(t)}{dt} - \frac{dY(t)}{dt}\right) + \lambda_0 e(t) = 0 \tag{25}$$

Solving $\frac{d^2Y(t)}{dt^2}$ from Eq. (25) and replacing it in Eq. (21):

$$U_c(t) = \frac{1}{K}\left[\tau\frac{d^2R(t)}{dt^2} + \tau\lambda_1\frac{dR(t)}{dt} + (1-\tau\lambda_1)\frac{dY(t)}{dt} + \tau\lambda_0 e(t)\right] \quad (26)$$

After quantitative analysis, we determined that it is feasible to discard the derivatives of the reference without causing any effects on the controller's performance, and providing greater simplicity in the control law.

The continuous part of SMC-IS is expressed in Eq. (27).

$$U_c(t) = \frac{1}{K}\left[(1-\tau\lambda_1)\frac{dY(t)}{dt} + \tau\lambda_0 e(t)\right] \quad (27)$$

Then, the full expression of SMC-IS is presented as follows

$$U(t) = \frac{1}{K}\left[(1-\tau\lambda_1)\frac{dY(t)}{dt} + \tau\lambda_0 e(t)\right] + K_d\frac{S(t)}{|S(t)|+\delta} \quad (28)$$

and,

$$S(t) = sign(K)\left(-\frac{dY(t)}{dt} + \lambda_1 e(t) + \lambda_0\int_0^t e(t)\,dt\right) \quad (29)$$

SMC-IS parameters (K_d, δ, λ_1 and λ_0) were calculated using the Particle Swarm Optimization algorithm. The objective function is a weighted value of the performance indexes (ISE and TVu), and the transient characteristics (Overshoot and Settling Time), in order to improve the system's response without disregarding the controller output.

5 Simulation Results

5.1 ECG Tracking and Regulation Test for Zeeman's Heartbeat Model Using SMC-FO-PSO

In order to compute the model's gain and time constant, a ten percent change to the input is introduced. The obtained parameters, using Smith and Corripio's method [16], are: $K = 1.323$ and $\tau = 0.408$; thus, using Eq. (19), $\lambda = 3.243$. As stated in previous sections; K_d and δ, will be obtained through PSO algorithm. The objective function J is presented in Eq. (30):

$$J = \frac{0.8ISE + 0.2TVu}{ISE + TVu} \quad (30)$$

The used parameters for the PSO algorithm are as follows:

- Number of particles: 10
- Number of iterations: 20

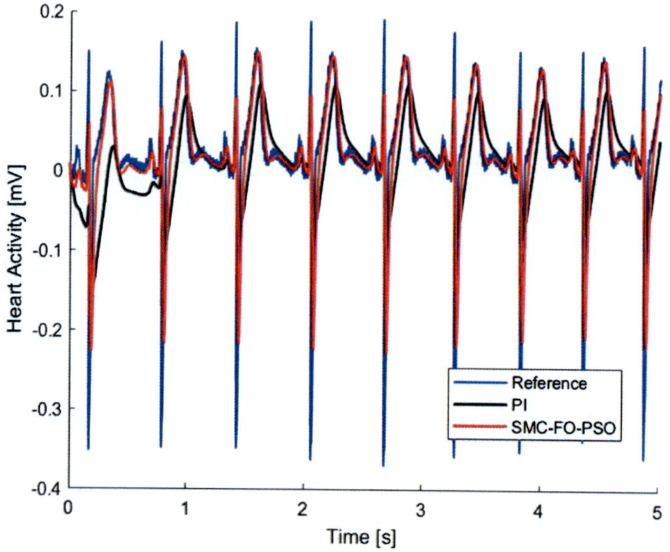

Fig. 2. System's response to a tracking test

After executing the algorithm, the results are as follows: $K_d = 74.99$ and $\delta = 2.61$. The model and the proposed controller were tested using a signal reference of an electrocardiogram of 80 [bpm] heart-rate, obtained from previously published work [1]. The proposed control scheme was then compared to a PI controller using Internal Model Control (IMC) tuning equations found in [15]. PI gains are: $K = 7.54$ and $Ti = 0.408$. The system's output is shown in Fig. 2. The proposed controller is capable of tracking the complex reference signal of an ECG; however, it has difficulties reaching the maximum values of the ECG. In contrast, the PI controller produces a similar output to the reference, but its response is worse compared to the SMC-FO-PSO.

The control signal is presented in Fig. 3. Even though the parameters were optimized in order to have a minimum TVu, the control signal from SMC-FO-PSO still presents some chattering due to the constant changes in the reference signal. The PI control signal has a smoother response with little chattering, but it has the disadvantage that the tracking capability is not accurate, which is represented by a higher ISE.

In order to compare both controllers quantitatively, performance indexes ISE and TVu are presented. The indexes for SMC-FO-PSO are: $ISE = 0.014$ and $TVu = 4.149$; for the PI the results are: $ISE = 0.028$ and $TVu = 4.605$. ISE results show that SMC-FO-PSO has a superior tracking capability, which is desirable in the proposed system since the controller should be able to follow a healthy heart activity. TVu results show that PI exerts less stress on the final control element.

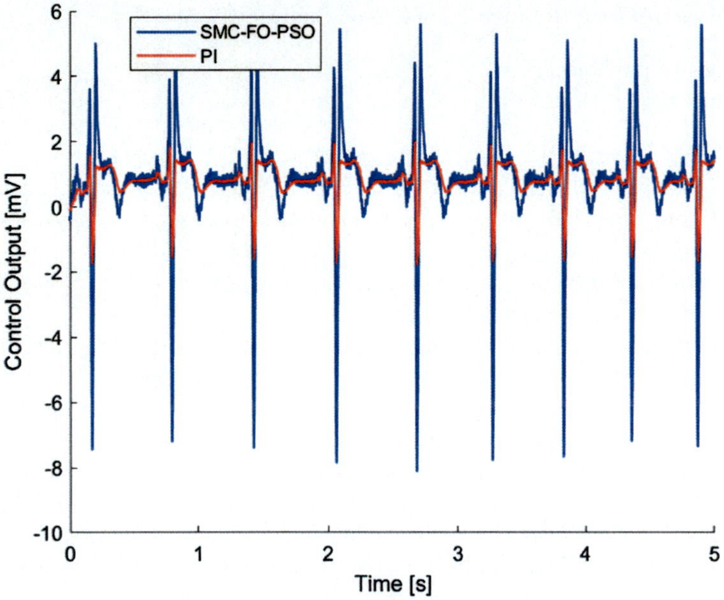

Fig. 3. Controller's output for the tracking test

5.2 Heart Rate Tracking Test Using SMC-IS-PSO

The Heart Rate model is an integrating second-order process. However, the SMC-IS-PSO's design demands a reduced-order integrating system; therefore, the methodology proposed by Henriquez and Martinez [9] is adopted to obtain the approximate integrating first-order model as shown in Eq. (31).

$$G_m(s) = \frac{7.857}{s(0.1665s+1)} \tag{31}$$

SMC-IS-PSO will be subjected to static and varying heart rates in order to determine the controller's performance for tracking purposes. In addition, a quantitative analysis will be carried out in relation to a PID controller which uses the IMC tuning method described by Chien and Fruehauf [4,15].

Static Heart-Rates: SMC-IS-PSO parameters (K_d, δ, λ_1 and λ_0) were found by applying the PSO metaheuristic algorithm where the objective function J is detailed in Eq. (32).

$$J = 0.6\ SettlingTime + 0.3\ Overshoot + 0.05\ ISE + 0.05\ TVu \tag{32}$$

The used values for PSO algorithm are as follows:

- Number of particles: 50
- Number of iterations: 30

SMC-IS-PSO and PID parameters are presented in Table 1.

The system's responses are shown in Figs. 4 and 5, where it is clear that the proposed controller has transient characteristics that significantly benefit the cardiac model's response [12]. Compared to PID, SMC-IS-PSO does not present a representative overshoot which implies improving the patient's condition in case of using an artificial pacemaker.

PID controller response's overshoot may be associated with an increase in patients' heart rate, which in medical terms would be considered tachycardia (above 100 [bpm]) [7]. Although this increase is almost instantaneous, about 1.5 s, (see Fig. 5a) until it adjusts to the static heart rate reference, it is likely to generate a stabbing pain in the patient's chest area. On the other hand, SMC-IS-PSO does not present this sudden alteration, and no pain will be inflicted on the patient.

Considering the control outputs (see Figs. 4b, and 5b), the PID controller has a magnitude of at least 5 times larger than the proposed controller. Therefore, SMC-IS-PSO has a much smoother control output and is a very promising option to implement in an artificial pacemaker prototype.

Table 1. Controllers tuning

Controller	K_c	T_i	T_d	K_d	δ	λ_1	λ_0
PID-Chien	0.9555	0.8325	0.0666	–	–	–	–
SMC-IS-PSO	–	–	–	0.0015	83.8994	10.2338	46.1590

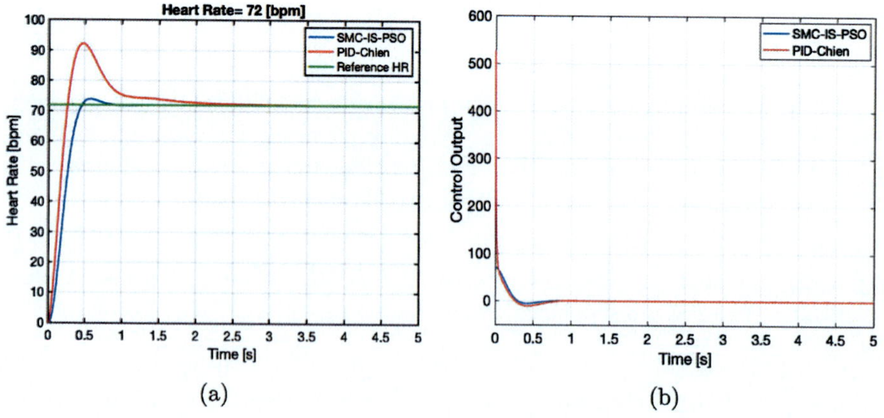

Fig. 4. Performance comparison: (a) System's response to a tracking test, and (b) Controllers' output to a tracking test at 72 [bpm]

Performance indexes and transient characteristics of static heart rates are presented in Table 2.

Table 2. Performance indexes and transient characteristics of static heart rates

Static heart-rate	Controller	ISE	TVu	Overshoot [%]	Ts [s]
72 [bpm]	SMC-IS-PSO	885.4	11.93	2.743	0.6781
	PID-Chien	729.7	17.71	28.073	1.6983
83 [bpm]	SMC-IS-PSO	1177	13.19	2.752	0.6783
	PID-Chien	969.7	19.52	28.148	1.6982

The quantitative analysis highlights that SMC-IS-PSO has the lowest magnitude values in most cases, which imply that the proposed controller has the best characteristics for the Heart-rate model's response [12].

Varying Heart-Rates: For this case, a reference with varying heart rates ranging from 80 to 92 [bpm] is used to verify if the proposed controller can be adjusted to a variable reference.

Since the reference is different from a step input, the system characteristics are modified and therefore the tuning is affected. A new objective function J is described in Eq. (33) in order to obtain the best performance for SMC-IS-PSO on a varying heart-rate setpoint.

$$J = 0.9\ ISE + 0.1\ TVu \tag{33}$$

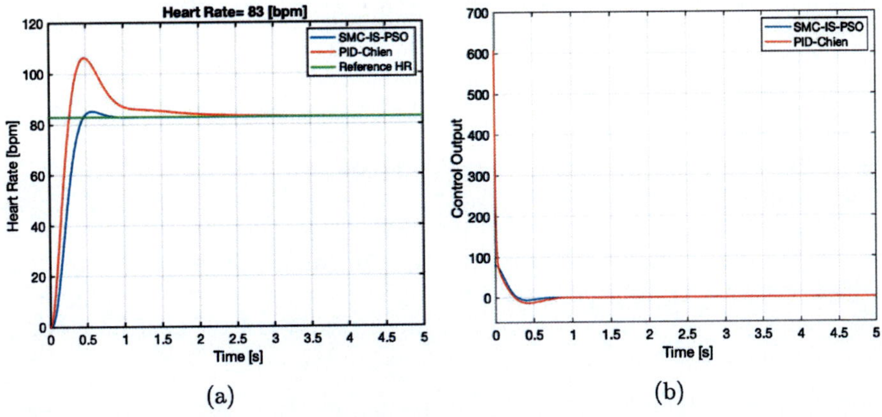

Fig. 5. Performance comparison: (a) System's response to a tracking test, and (b) Controllers' output to a tracking test at 83 [bpm]

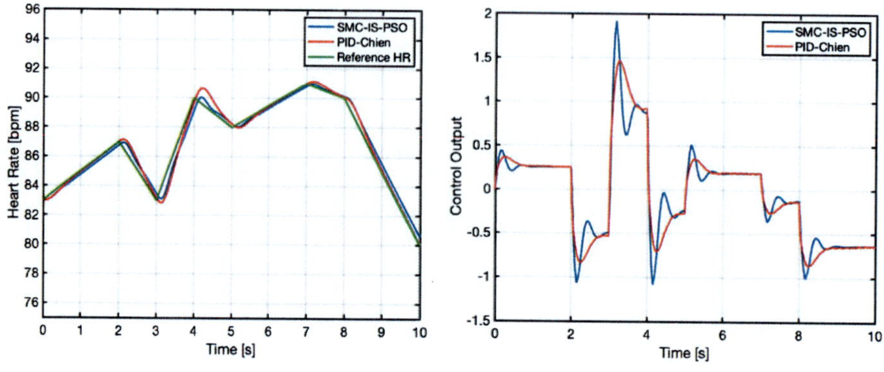

(a) (b)

Fig. 6. Performance comparison: (a) System's response to a tracking test, and (b) Controllers' output to a tracking test at varying heart-rates (80–92 [bpm])

The used values for PSO algorithm are as follows:

- Number of particles: 20
- Number of iterations: 10

SMC-IS-PSO and PID parameters are presented in Table 3.

Table 3. Controllers tuning at varying heart-rates

Controller	K_c	T_i	T_d	K_d	δ	λ_1	λ_0
PID-Chien	0.9555	0.8325	0.0666	–	–	–	–
SMC-IS-PSO	–	–	–	0.0021	926.3078	11.9769	100.2057

The obtained results are described in Fig. 6. The proposed controller is the one that best adjusts to varying heart rates, being a promising alternative for the development of an artificial pacemaker prototype.

Performance indexes of varying heart rates for SMC-IS-PSO are ISE = 1.922 and TVu = 13.19 and for Chien's PID are ISE = 2.188 and TVu = 19.52.

The quantitative analysis based on the performance indexes emphasizes that SMC-IS-PSO is the controller with the best characteristics for varying heart rates, denoting the robustness of Sliding Mode theory and the controller's optimization through the execution of the PSO algorithm.

6 Conclusions

Since heart systems have complex dynamics and a highly variable reference such as ECG, it was desirable that the proposed controllers had a better performance than the traditional ones, which was fulfilled for tracking purposes with satisfactory results.

The tracking capabilities of SMC-FO-PSO are superior when compared to a traditional PID controller; this is quantitatively proven by the ISE indexes computed from the obtained results. A lower ISE index represents a minor error, resulting in a response that better tracks the ECG set-point.

Even though the tuning parameters for the SMC-FO controller were obtained from the PSO algorithm, some control chattering can be observed from the controller's output graph. Control chattering can result in final-control element stress, which in turn can yield a lower lifespan of the pacemaker. This control behaviour can be improved with manual tuning of the δ parameter.

When comparing the heart rate tracking performance, SMC-IS-PSO yields a better response with minimal overshoot. Minimizing overshoot is critical in order to reduce tachycardia risk in patients using a pacemaker. An important metric when analyzing the controller's performance is the output signal; SMC-IS-PSO gives a smoother controller output, reducing final control element strain.

When applied to a variable heart rate set-point, SMC-IS-PSO adjusts better to the reference signal compared to the PID controller. Based on the ISE index, the error is lower when SMC-IS-PSO is used.

References

1. American heart Association: American heart association ecg database usb. https://www.ecri.org/american-heart-association-ecg-database-usb
2. Bonyadi, M.R., Michalewicz, Z.: Particle swarm optimization for single objective continuous space problems: a review. Evol. Comput. **25**(1), 1–54 (2017)
3. Camacho, O., Smith, C.: Sliding mode control: an approach to regulate nonlinear chemical processes. ISA Trans. **39**, 205–218 (2000). https://doi.org/10.1016/S0019-0578(99)00043-9
4. Chien, I.L., Fruehauf, P.S.: Consider IMC tuning to improve controller performance. Chem. Eng. Prog. **86**(10), 33–41 (1990)
5. Enamorado, A., Garcia, I., Gonzalez, M., Goro, G.: Clinical and epidemiological characterization of patients with permanent pacemaker implants. Revista de Ciencias Médicas de Pinar del Río 24 (2020)
6. Gomez, R., Carrasco, S.: Pacemaker carrier patient in the primary health care consultation. Medicina de Familia. Semergen **31**, 365–369 (2005)
7. Gopinathannair, R., Olshansky, B.: Management of tachycardia. F1000Prime Reports 7 (2015). https://doi.org/10.12703/P7-60
8. Govind, A., Sekhar, A.: A novel design of an adaptive PID controller for cardiac pacemaker (2014)
9. Henriquez, J., Martinez, W.: Identificación y sintonización de controladores pid para procesos de integración (2019)
10. Hung, J., Gao, W., Hung, J.: Variable structure control: a survey. IEEE Trans. Ind. Electron. **40**(1), 2–22 (1993). https://doi.org/10.1109/41.184817
11. Institute, T.H.: Pacemakers. https://www.texasheart.org/heart-health/heart-information-center/topics/pacemakers/

12. Khan, P., Khan, Y., Kumar, S.: Tracking and stabilization of heart-rate using pacemaker with fof-pid controller in secured medical cyber-physical system. In: 2020 International Conference on COMmunication Systems NETworkS (COMSNETS), pp. 658–661 (2020). https://doi.org/10.1109/COMSNETS48256.2020.9027302
13. Kumar, S.M.G., Sivasankar, R., Radhakrishnan, T.K., Dharmalingam, V., Anantharaman, N.: Particle swarm optimization technique based design of pi controller for a real-time non-linear process. Instrument. Sci. Technol. **36**(5), 525–542 (2008). https://doi.org/10.1080/10739140802234980
14. Priyadarshi, P.A., Kannaiyan, S.: Estimation of human heart activity using ensemble kalman filter. Sensors Transd. **209**, 90–96 (2017)
15. Seborg, D.E., Edgar, T.F., Mellichamp, D.A., Doyle, F.J.: Process Dynamics and Control, 3rd edn. John Wiley and Sons Inc., Hoboken (2011)
16. Smith, C., Corripio, A.: Principles and Practice of Automatic Process Control. John Wiley & Sons, Hoboken (2005)
17. Talange, D., Laware, A., Bandal, V.: Development of an internal model sliding mode controller for cascade control system. In: 2015 International Conference on Energy Systems and Applications, pp. 51–56 (2015). https://doi.org/10.1109/ICESA.2015.7503312
18. Thanom, W., Loh, R.: Observer-based nonlinear feedback controls for heartbeat ECG tracking systems. Intell. Control Autom. **03**, 251 (2012). https://doi.org/10.4236/ica.2012.33029
19. Utkin, V.: Variable structure systems with sliding modes. IEEE Trans. Autom. Control **22**(2), 212–222 (1977). https://doi.org/10.1109/TAC.1977.1101446
20. Vaidyanathan, S., Lien, C.-H. (eds.): Applications of Sliding Mode Control in Science and Engineering. SCI, vol. 709. Springer, Cham (2017). https://doi.org/10.1007/978-3-319-55598-0
21. Vincent, A., Nersisson, R.: Particle swarm optimization based PID controller tuning for level control of two tank system. In: IOP Conference Series: Materials Science and Engineering, vol. 263, p. 052001 (2017). https://doi.org/10.1088/1757-899X/263/5/052001
22. Yadav, J., Rani, A., Garg, G.: Intelligent heart rate controller for cardiac pacemaker. Int. J. Comput. Appl. **36**(7), 22–29 (2011)
23. Zeeman, E.: Differential equations for the heartbeat and nerve impulse††ams (mos) 1970 subject classification: 35f99. In: Peixoto, M. (ed.) Dynamical Systems, pp. 683–741. Academic Press (1973). https://doi.org/10.1016/B978-0-12-550350-1.50055-2

Design, Simulation, and Implementation of an Artificial Pancreas Prototype for Virtual Patients with Type 1 Diabetes Applying SMC Controller with Anticipated Carbohydrate Information

Stefany Villarreal[1], Diego Lombeida[1], Jenny Haro[1], and Oscar Camacho[1,2](✉)

[1] Departamento de Automatización y Control Industrial, Escuela Politécnica Nacional, Quito, Ecuador
{stefany.villarreal,diego.lombeida,jenny.haro, oscar.camacho}@epn.edu.ec

[2] Colegio de Ciencias e Ingenierías "El Politécnico", Universidad San Francisco de Quito USFQ, Quito 170157, Ecuador
ocamacho@usfq.edu.ec

Abstract. This paper presents the simulation, design, and implementation of an insulin pump prototype for virtual patients with type 1 diabetes using an SMC controller, validated with the UVA/PADOVA simulator. The results obtained with an average adult patient with a scenario of 5 meals for one day offer 88.88% time in range.

Keywords: Diabetes · Exogenous insulin · Glucose · Carbohydrates · Prototype · Insulin pump

1 Introduction

Diabetes is a disease that increases annually worldwide, affecting the quality of life of people who suffer from it and their close relatives. An inadequate diet produces this and a lack of physical activity, and an autoimmune reaction of the human body, killing the beta cells responsible for producing insulin [1].

The treatment for this disease is based on changing the type of diet, which is challenging to comply with due to nutritional diets. Patients undergo insulin infusions external to the body, which are placed using daily punctures, generating trauma. In addition, manual insulin infusions present difficulties in glucose regulation since there is no monitoring of it throughout the day, producing high and low glucose concentrations in patients that can even cause death and can occur suddenly [2].

Previous to this work, at the Escuela Politécnica Nacional, three theses corresponding to the authors [3, 4] y [5] have been carried out. The design and comparison of controllers based on different mathematical glucose-insulin models are supported only

M. Botto-Tobar et al. (Eds.): *Latest Advances in Electrical Engineering, and Electronics*, LNEE 933, pp. 115–128, 2022.
https://doi.org/10.1007/978-3-031-08942-8_9

by simulations with the Simulink/MATLAB software. In addition, a virtual patient with type 1 diabetes has been developed [6].

This paper proposed to make an initial artificial pancreas prototype, which could be used in the future for the patient to lead a better quality of life. Using Hovorka's first, a simulated virtual patient was implemented in Simulink/Matlab. Next, an SMC controller was designed to regulate the virtual patient's glucose, validated with UVA/PADOVA simulator, the unique simulator approved by the FDA to perform preclinical testing of virtual patients with type 1 diabetes [7]. In the end, the insulin pump prototype was developed using the SMC driver. The pump was tested with Hovorka's virtual patient, and the pump characteristics were validated with the UVA/PADOVA simulator.

This work is divided as follows: section two describes the materials and methods, in section three, the results are shown, and finally, some conclusions are presented.

2 Materials and Methods

This section describes some fundamentals about diabetes, a glucose-insulin mathematical model used to design an SMC controller, and finally, the physical prototype design.

2.1 Theoretical Foundation

The artificial pancreas is a device that controls glucose levels in people without delivering insulin. It is composed of a sensor that measures the glucose levels in the interstitial fluid of the pancreas, the control algorithm for the insulin delivery pump, and a cannula linked to a subcutaneous catheter that connects the pump to the patient [8].

People with Type 1 diabetes must inject insulin manually to keep glucose levels within the normoglycemic limits (70[mg/dl]-180[mg/dl]); if the patient received an incorrect amount, it could lead to complicated cases of:

Hypoglycemia (<70[mg/dl]). It occurs when plasma glucose is very low due to not ingesting enough foods with glucose content, the infusion of an excessive amount of exogenous insulin, or the excessive amount of physical exercise performed. Symptoms are dizziness, sweating, feeling confused [8].

Hyperglycemic values (>180[mg/dl]). It is due to low insulin production, poor insulin action, insulin resistance, and eating foods high in CHO. The symptoms are tiredness, weight loss, excessive thirst, headaches, and frequent urine [8].

Therefore, the injections can be avoided with the use of a semi-automatic artificial pancreas in which the patient enters the number of carbohydrates ingested, and the controller will oversee injecting the amount of basal insulin needed throughout the day and the amount of insulin bolus in the three daily meals, improving the quality of life of patients [8].

Hovorka's Model. It is a model that describes the interaction between glucose and insulin, which is made up of two compartmental sub-models that describe the kinetics of insulin and subcutaneous glucose. A third sub-model comprises two compartments representing glucose absorption in the gastrointestinal tract [9].

2.2 Describing Modeling Equations

Virtual Patient Model. A virtual patient with type 1 diabetes was simulated in Simulink/Matlab software, the Hovorka's model [9] was used, with the following equations.

Dynamics of Blood Glucose. It refers to the amount of glucose measured in interstitial fluid.

$$\frac{dG(t)}{dt} = -S_I(X(t) - X_b) + U_M(t) - K(G(t) - G_b) \tag{1}$$

where:

$G(t)$ Glucose concentration in the blood. [mmol/l]
S_I Insulin sensitivity. [mmol/l/min] per [mU/l]
X_b Effective basal insulin concentration, which should have a constant value. [mU/l]
G_b Basal glucose level. [mmol/l]
K Fractional range of glucose self-regulation. [1/min]

Insulin absorption and action model. In Eqs. (2) and (3), the absorption of insulin passes through two compartments before being used by the body, and in Eq. (4), the action of insulin is described.

$$\frac{dx_1(t)}{dt} = -\frac{1}{t_{max,IA}}x_1(t) + \frac{u_I(t)}{60} \tag{2}$$

$$\frac{dx_2(t)}{dt} = \frac{1}{t_{max,IA}}(x1(t) - x_2(t)) \tag{3}$$

$$\mathrm{X(t)} = \frac{1000.x_2(t)}{t_{max,IA}MCR_I W} \tag{4}$$

where

$x_1(t)$ Amount of effective insulin in the first compartment. [U]
$x_2(t)$ Amount of effective insulin in the second compartment. [U]
$u_I(t)$ Insulin infusion rate at time t. [U/h]
$t_{max,IA}$ Time to maximum insulin concentration. [min]
$X(t)$ Effective insulin concentration. [mU/l]
W Weight. [Kg]
$MCRI$ Effective insulin rate. [l/kg/min]

Food Absorption Dynamics Model. Describes, employing Eqs. (5), (6), and (7), the process that is carried out with the ingestion of CHO, which is the primary source of glucose, which initially passes from the stomach to the small intestine and then it reaches the interstitial fluid.

$$\frac{da_1(t)}{dt} = -\frac{1}{t_{max,G}}a_1(t) + \delta_{t_j}(t)u_G(t_j) \tag{5}$$

$$\frac{da_2(t)}{dt} = \frac{1}{t_{max,G}}(a_1(t) - a_2(t)) \tag{6}$$

$$U_M(t) = \frac{5,556.A_G a_2(t)}{t_{max,G} V_G W} \tag{7}$$

where

$a_1(t)$ Amount of carbohydrates in the first food absorption compartment. [g]
$a_2(t)$ Amount of carbohydrates in the second food absorption compartment. [U]
$u_G(tj)$ Amount of carbohydrates consumed in time tj. [g]
$t_{max,G}$ Time of maximum onset of glucose variation. [min]
AG Fractional bioavailability [unitless]
VG Plasma glucose pool size. [1/Kg]
$UM(t)$ Rate of intestinal carbohydrate absorption with unit converted to rate of change of glucose concentration. [mmol/l/min]

2.3 Sliding Mode Controller

An SMC controller was developed because it is robust to perturbations and parametric variations; it allows patient parameter changes.

Camacho and Smith control law [10] is used for the controller design, consisting of the Eq. (8) and a PID surface (9).

$$U(t) = \left(\frac{t_0\tau}{K}\right)\left[\frac{X(t)}{t_0\tau} + \lambda_0 e(t)\right] + K_D\frac{S(t)}{|S(t)| + \delta} \tag{8}$$

$$s(t) = -\frac{dy(t)}{dt} + \lambda_1 e(t) + \int \lambda_0 e(t)dt \tag{9}$$

where:

$$K_D = \frac{0.51}{|K|}\left[\frac{\tau}{t_0}\right]^{0.76} \tag{10}$$

$$\lambda_1 = \frac{t_0 + \tau}{t_0\tau} \tag{11}$$

$$\lambda_0 \leq \frac{\lambda_1^2}{4} \tag{12}$$

$$\delta = 0.68 + 0.12(KK_D\lambda_1) \tag{13}$$

2.4 Pancreas Artificial Prototype Control Loop

The control loop is divided into two parts. The first one considers software, and the other one is hardware. Figure 1 illustrates the control loop with the prototype, and in Fig. 2 are the control loop's structure inside the prototype.

Software where are a virtual patient; designed with Hovorka's Model; and virtual sensor; based on [11]. the information is sent to the prototype by serial communication. To validate the controller and pump, UVA/PADOVA are used.

Hardware, here is the prototype that contains the glucose setpoint and the glucose information received from the computer. The error signal is obtained that enters the controller. The carbohydrate advance allows manipulating the engine and reporting the amount of insulin that enters the virtual patient.

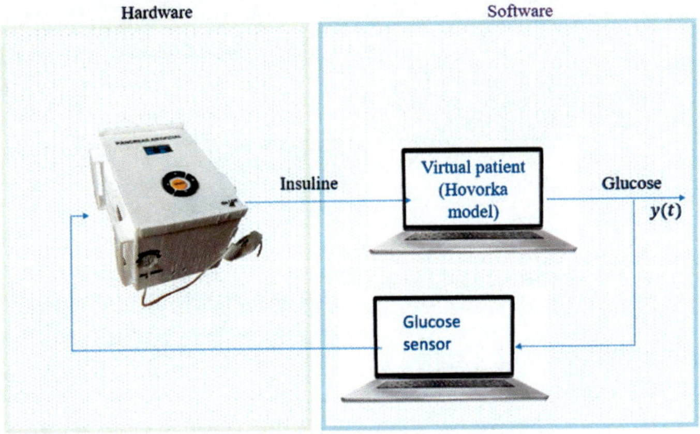

Fig. 1. Control loop with a prototype

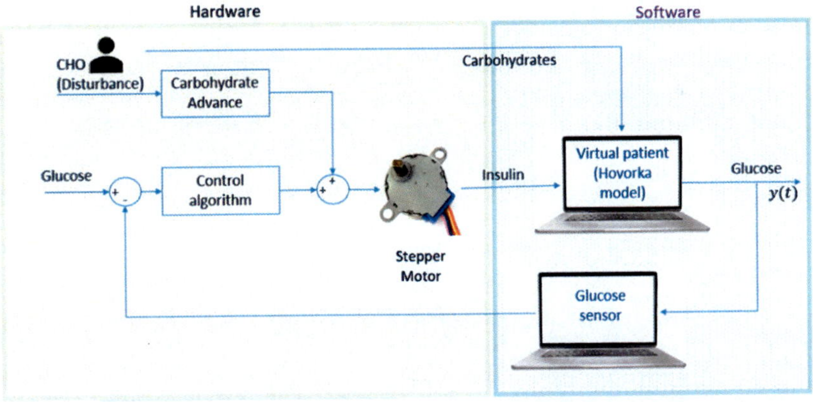

Fig. 2. Control loop inside of the prototype

2.5 The Graphical Interface (HMI)

The graphical interface was developed to observe the glucose results of virtual patients 1, 4, and average with type one diabetes of Hovorka's model, applying the developed controller; it also allows the visualization of the prototype results and validation of validation themselves.

It was implemented in Matlab's App Designer and Simulink based on the ISO 13485 [12] standard used for biomedical devices (Fig. 3 and Fig. 4).

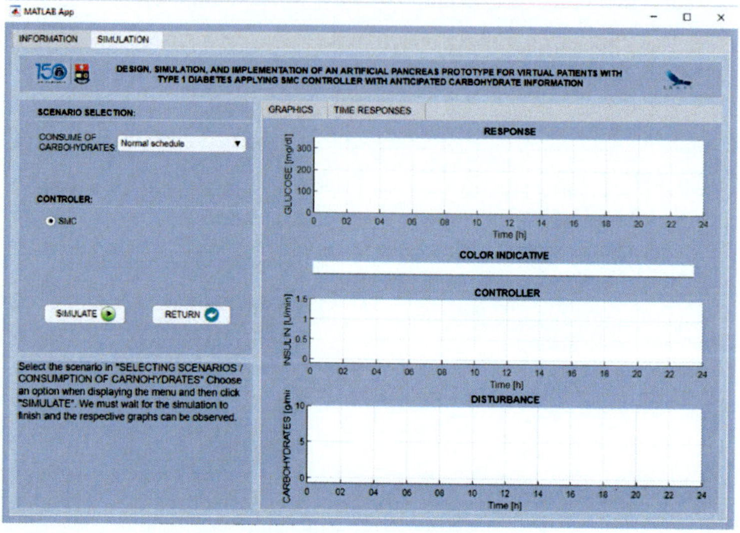

Fig. 3. HMI to observe the prototype results

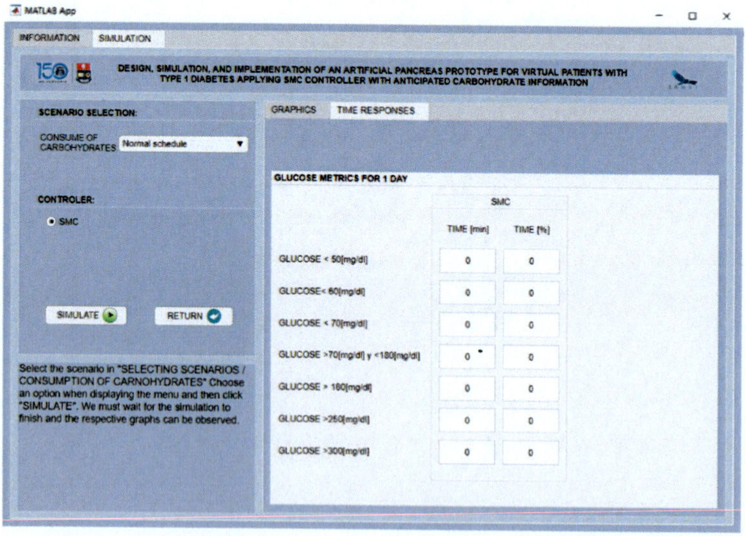

Fig. 4. HMI with metrics for one day

2.6 Electronic Devices for Assembly

For the implementation of the prototype, the following elements were considered:

- Stepper motor 28BYJ-45 unipolar 5V with reduction box [13].
- ESP32S microcontroller in charge of complying with the processes involved in the prototype of the artificial pancreas in addition to having the number of pins required for the connection of peripherals [14].
- OLED SH1106 screen for displaying information on the prototype [15].
- Lithium battery LGABC41865 of 3.7V and 2750mAh to power the motor [16].
- TP4056 Lithium Battery Charger to store energy back into the selected battery.
- MT3608 step-up DC/DC converter to increase the voltage from 3.7 V to 5 V required for motor operation [17].
- Driver ULN2003 to amplify the current leaving the microcontroller and entering the motor [18].
- Quick-Set infusion set to transport insulin from the insulin pump into a person's abdomen [19].
- 3 ml MiniMed insulin reservoir, which stores insulin [20].

2.7 Injection System

For insulin injection, a rack and pinion system is used, where a pinion is attached to the motor shaft, which allows linear movements to be made when moving through a rack. This system supports Fig. 5 developed in AutoCAD inspired by [4] with the insulin reservoir.

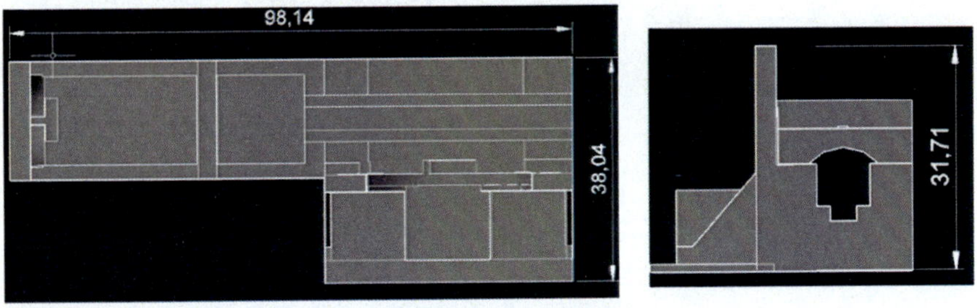

Fig. 5. Injection system

The minimum amount of insulin supplied by the prototype pump is 0.059U; this value was obtained by applying Eqs. (14), (15), and (16) [21].

$$d = \frac{Z}{n} \tag{14}$$

$$D_v = \frac{d}{p} \tag{15}$$

$$V = \pi r^2 D_v \tag{16}$$

where:

d The rack carries out the movement in one turn of the pinion.
Z Number of pinion teeth
n Number of zipper teeth in 1 [cm].
D_v Distance traveled by the rack in 1 step of the motor.
p The number of motor steps to complete 1 turn.
V Amount of insulin obtained with 1 step of the motor.
r Cylinder radius.

2.8 Printed Circuit Board (PCB)

The maximum current circulating through the circuit is 800 mA, so a 0.6 mm track width is taken following the IPC 2220 standard [22]. This circuit has two voltage levels, 3.7 V and 5 V. The schematic circuit and PCB were designed in Altium software.

2.9 Prototype Case

It was designed in the AUTOCAD software; there are the elements necessary for the pump to dose the amount of insulin required to the human body.

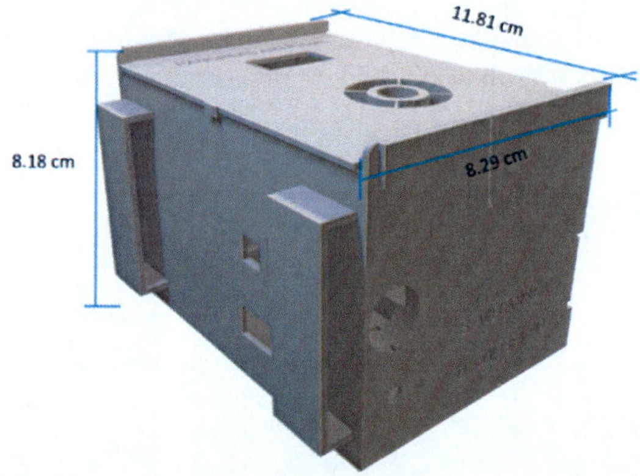

Fig. 6. Prototype case

The dimensions of the prototype can be seen in Fig. 6, which could be reduced in future improvements of the same.

3 Results

This section divides the results into four parts: simulation, controller validation, prototype, and pump validation.

3.1 Controller Simulation with Hovorka's Patient Model

The carbohydrate consumption scenario chosen to analyze the results obtained, With the average patient of Hovorka's model, is detailed in Table 1 because it is very similar to what a person eats on a typical day. This scenario is composed of two meals and two snacks.

Table 1. Carbohydrates consumed.

Foods	Carbohydrates [gm]	Time
Breakfast	40	07:00
Snack	25	10:00
Lunch	60	13:00
Snack	30	16:00
Diner	30	19:00
Total	185	

The response obtained with the controller with the test scenario can be seen in Fig. 7 b); it presents two cases of hyperglycemia, one from 8:30 a.m. to 9:30 a.m. and the other from 14:00 to 16:00 without causing hypoglycemia and a time in a range of 86.46% for a day.

3.2 Controller Validation with UVA/PADOVA Simulator

For this test, five adult patients (patient 1, patient 3, patient 4, patient 5, patient 10) were randomly selected, plus the population patient. The performance for the controller was verified, analyzed using the control variability table (CVG), which establishes by colors and zones, how good or bad a controller keeps glucose values within normoglycemia. Later, to proceed to implement it in hardware. The black dots located on the CVG matrix correspond to the maximum and minimum glucose levels obtained for a patient in 24 h [7].

a)

b)

c)

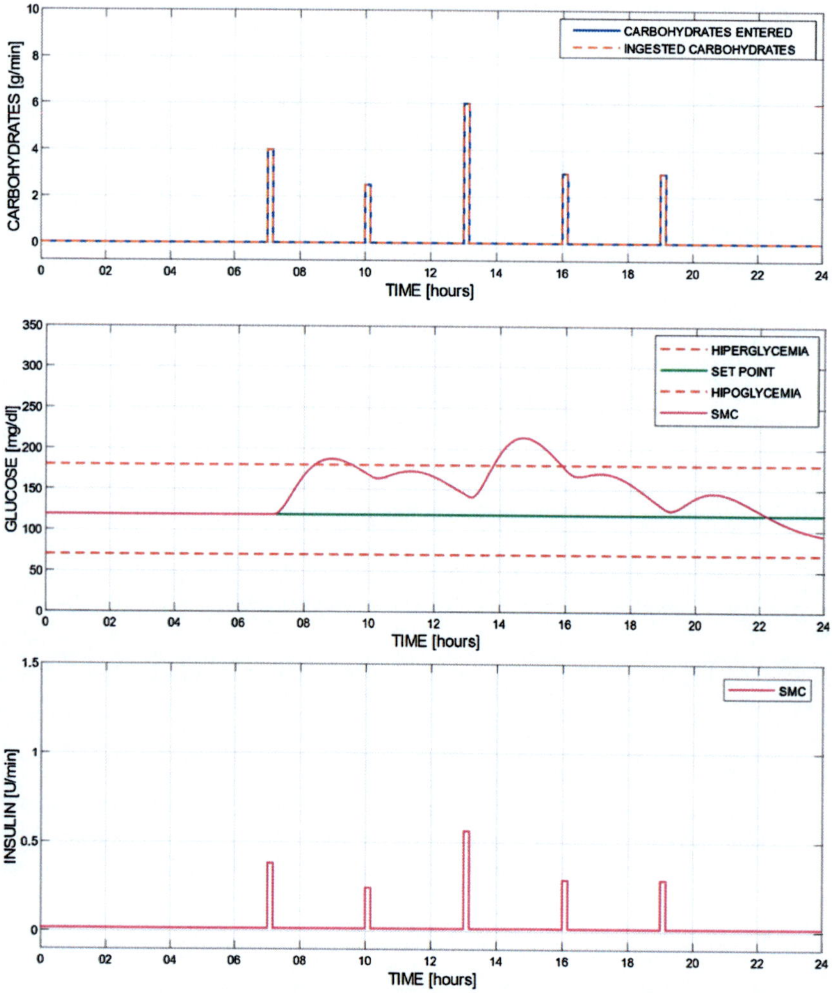

Fig. 7. The response obtained from the simulation for the test scenario: a) CHO intake; b) Blood glucose concentration; c) insulin infusion.

As can be seen in the results obtained with the UVA/PADOVA simulator in Fig. 8, the SMC controller ensures a time in range without hypoglycemic episodes.

According to the results obtained, it was deduced that the SMC controller is suitable for use in implementing the prototype since it keeps the patients in zone B of the CVG matrix of the UVA/PADOVA simulator.

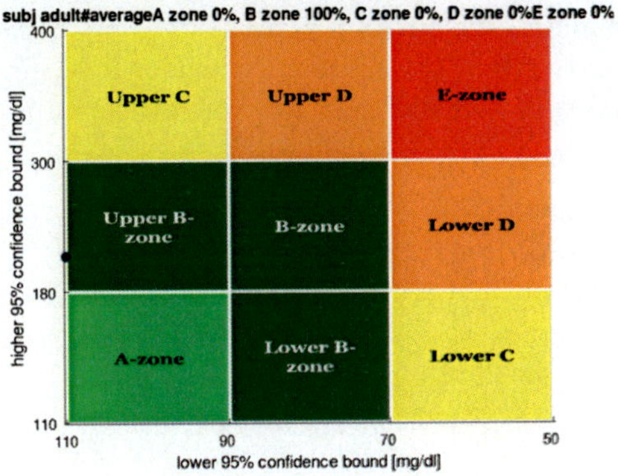

Fig. 8. CVG for the average adult patient of UVA/PADOVA

3.3 Prototype Results

The printed circuit was made in Bakelite by routing in a CNC machine which contains the elements selected according to the requirements.

The prototype was built with a 3D printer with PLA filament (Fig. 9); its weight is 290 g. It contains the mainboard, injection system, screen, and buttons inside the mainboard.

Fig. 9. Insulin drum prototype implemented

The capacity required by the circuit is 1452.28 mAh per day; considering that the battery has a capacity of 2200 mAh, an autonomy of 1 to 2 days is obtained.

The result with the same scenario in Table 1 with the controller in the prototype is:

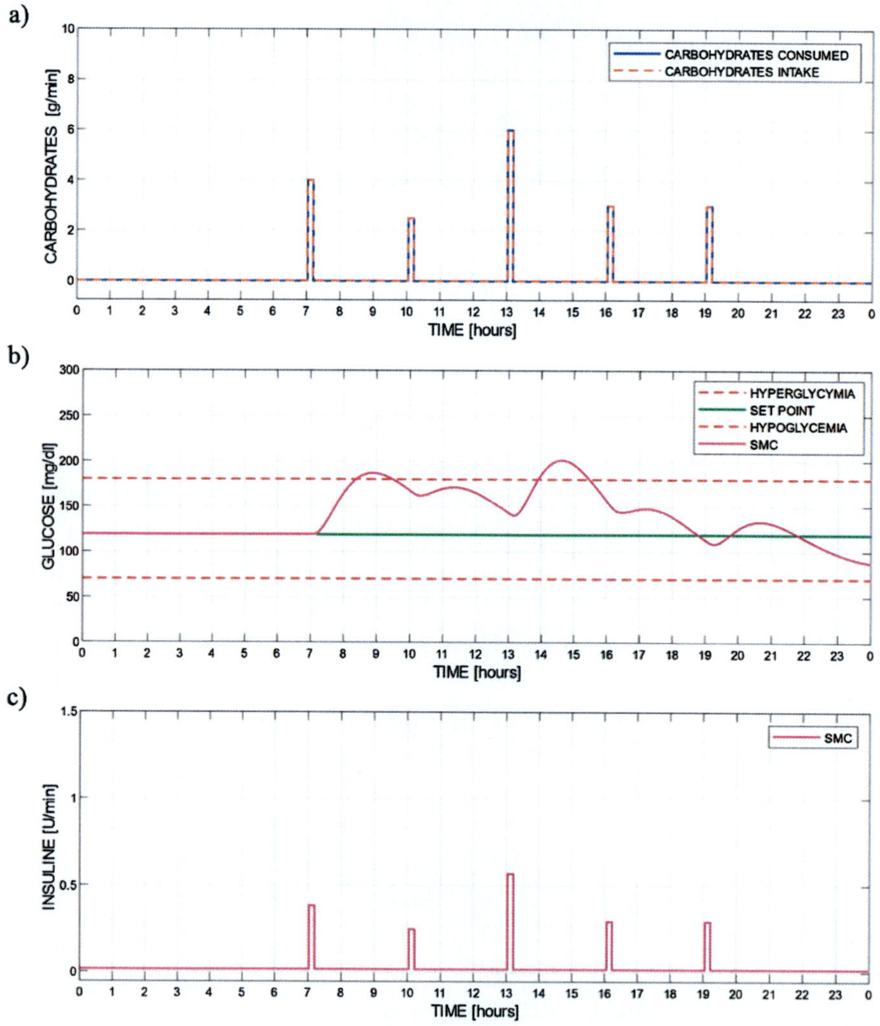

Fig. 10. The response obtained with the prototype: a) CHO intake; b) Blood glucose concentration; c) Insulin infusion.

In Fig. 10 b) the glucose response obtained when we consider the five daily meals for a patient with type 1 diabetes is observed, in which two cases of hyperglycemia occur after breakfast and lunch; however, your maximum glucose value is 200 [mg/dl], which is below the limit recommended by [1].

As mentioned above, this scenario is the one that comes closest to the type of diet that a patient with this disorder should take. As can be seen, the prototype keeps the patient within normoglycemia most of the day, obtaining a similar response to the one previously shown in simulation in Fig. 7 with 88.88% time in range.

A video of the simulation can be seen in [23].

3.4 Pump Validation

The ten virtual adult patients of the UVA/PADOVA are selected, and with the data of the developed prototype pump, the simulation is carried out with the SMC controller, obtaining the results shown in Fig. 11.

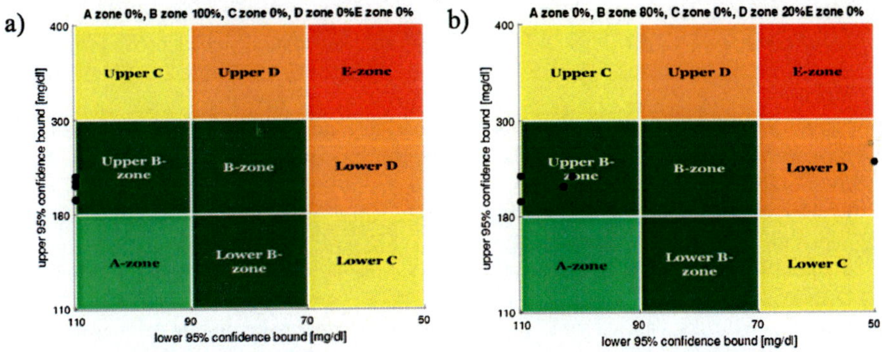

Fig. 11. CVG matrix of the UVA/PADOVA with the prototype made: a) patient 1 to 5; b) patient 6 to 10

The prototype pump performs adequate glucose control in 9 patients as these are within zone B, and only one patient is in zone D.

4 Conclusions

An SMC controller with anticipated carbohydrate information was designed using the Hovorka model as a virtual patient with type 1 diabetes. A scenario of 5 meals was applied for the simulation in 1 day. The SMC controller was validated using the CVG matrix of the UVA/PADOVA simulator, where the responses obtained are within zone B corresponding to a suitable controller.

An artificial insulin pump prototype for the pancreas is implemented, using cheap devices easily found in the Ecuadorian market; however, more expensive biomedical devices should be bought to improve the prototype.

Using the Hovorka model for the five meals a day scenario, similar responses were obtained when comparing the simulations with the hardware In the Loop, allowing us to validate the controller implemented within the embedded system.

The size of the device made can be reduced by using SMD elements and a smaller motor, increasing the price of the final product.

Acknowledgment. The authors thank the PIGR-19–17 Project of Escuela Politécnica Nacional for its support for performing this work.

References

1. Ministerio de Salud Pública: Diabetes mellitus tipo 1: diagnóstico y manejo. Quito (2019)
2. Bupa Ecuador: Retrieved from bupasalud (2020). https://www.bupasalud.com.ec/salud/diabetes-tipo-1
3. Alvaro Gualoto, A.E., Rivadeneira Vaca, S.F.: Diseño y simulación y comparación de controladores de concentración de glucosa en pacientes con diabetes tipo 1 (Bachelor's thesis, Quito) (2017)
4. Calupiña Moya, D.J., García Vásconez, A.P. Diseño, simulación y comparación de controladores clásicos y avanzados, aplicados al modelo de glucosa-insulina en el sistema de páncreas artificial para pacientes con diabetes tipo 1 (Bachelor's thesis, Quito) (2018)
5. Moreano Terán, B.J., Pumisacho Pinto, J.F.: Diseño, simulación y comparación de controladores basado en algebra lineal y control difuso, aplicado al modelo de glucosa-insulina del sistema de páncreas artificial para pacientes con diabetes tipo 1 (Bachelor's thesis, Quito) (2019))
6. Haro Bonilla, J.P.: Diseño y simulación de un paciente virtual con diabetes tipo 1 a partir de un modelo de glucosa–insulina para la simulación de un páncreas artificial (Bachelor's thesis, Quito) (2020)
7. Rosales, N.: Modelado y simulación de tecnologías para el tratamiento de la diabetes (Doctoral dissertation, Universidad Nacional de La Plata) (2020)
8. Medtronic Diabetes: Información básica de terapia con bomba de insulina. Medtronic Diabetes, California
9. Ruan, Y., Wilinska, M.E., Thabit, H., Hovorka, R.: Modeling day-to-day variability of glucose–insulin regulation over 12-week home use of closed-loop insulin delivery. IEEE Trans. Biomed. Eng. **64**(6), 1412–1419 (2016)
10. Camacho, O., Smith, C.A.: Sliding mode control: an approach to regulate nonlinear chemical processes. ISA Trans. **39**(2), 205–218 (2000)
11. Choleau, C., et al.: Calibration of a subcutaneous amperometric glucose sensor: Part 1. Effect of measurement uncertainties on the determination of sensor sensitivity and background current. Biosensors and Bioelectronics **17**(8), 641–646 (2002)
12. Abuhav, I.: ISO 13485: 2016: a complete guide to quality management in the medical device industry. CRC Press (2018)
13. Wotiom: Stepper motor Model: WS23-0150-30-4. Retrieved from www.wotiom.com
14. Ai-Thinker: ESP-32S Datasheet (2016). Retrieved from http://www.ai-thinker.com
15. Solomon Systech: SSD1306 (2008). Retrieved from http://www.solomon-systech.com/
16. NanJing Top Power ASIC Corp: TP4056 1A Standalone Linear Li-lon Battery Charger with Thermal Regulation in SOP-8. Retrieved from https://www.mikrocontroller.net/attachment/273612/TP4056.pdf
17. Aerosemi: MT3608. Retrieved from https://www.olimex.com/Products/Breadboarding/BB-PWR-3608/resources/MT3608.pdf
18. ST: ULN2001A-ULN2002A. Retrieved from https://www.st.com/
19. Medtronic Diabetes: Quick-set® | Medtronic. Retrieved from https://www.medtronicdiabeteslatino.com/productos/equipos-de-infusion/quick-set
20. Medtronic Diabetes: Sure-T® | Medtronic. Retrieved from https://www.medtronicdiabeteslatino.com/productos/equipos-de-infusion/sure-t
21. Cnice: (2021) Retrieved from http://concurso.cnice.mec.es/cnice2006/material107/mecanismos/mec_cremallera-pinon.htm
22. Al delta: Normas internacionales básicas para diseño de circuitos impresos PCB y productos electrónicos. Retrieved from www.aldelta.com.co
23. Ste fy: Bomba de insulina para pacientes con diabetes tipo 1, escenario con Breaks [Motion Picture] (2021). Retrieved from https://www.youtube.com/watch?v=lpA7tdCl7mM

Design and Characterization of a Wireless Illuminance Meter with IoT-Based Systems for Smart Lighting Applications

Ricardo Araguillin[1,2](✉), Angel Toapanta[1], Daniela Juiña[1], and Byron Silva[1]

[1] Instituto de Investigación Geológico y Energético IIGE, Quito, Ecuador
{angel.toapanta,daniela.juina,byron.silva}@geoenergia.gob.ec
[2] Grupo de Láser, Óptica de Materiales y Aplicaciones Electromagnéticas GLOmAe, Facultad de Ingeniería, Universidad de Buenos Aires, Buenos Aires, Argentina
raraguillin.ext@fi.uba.ar

Abstract. Healthcare areas require high quality levels of lighting, and visual comfort. The current reality prevents access to these areas, so the use of wireless measurement systems has become a necessity . In this work, we present the design and characterization of a wireless illuminance meter that uses a BH1750 ambient light sensor and an ESP32 processor that allow data to be sent in real-time through IoT. The illuminance meter characterization showed that the error sources such as background radiation, shot noise, thermal noise are small enough and do not affect the measurement system and that the linearity of the sensor is high. On the other hand, we verify that the main source of error is given by the amount of oblique light that reaches the sensor (cosine correction) which could significantly error. Finally, we demonstrate that with an appropriate adjustment factor a high-performance illuminance meter with low errors can be achieved.

Keywords: Lux Meter · Internet of Things (IoT) · Wireless Sensor · BH1750

1 Introduction

The development of the Internet of Things (IoT) has generated significant changes in human life and has become a technological paradigm. In this context, everything in the physical world that is of interest to humans will be able to be connected and offer services through the Internet [1]. IoT presents a new way of seeing and controlling things because it brings the development of new communication protocols and new information management methods that facilitate automation and control processes. Thus, with the rise of IoT devices, smart lighting systems have become more popular in homes and cities [2]. Lighting control systems must be able to control the appropriate amount of light depending on the environment in which people are in such a way that it does not affect their performance, health and safety during the development of their activities, for which these systems must have sensors (luxmeters) capable of transmitting measurements in

M. Botto-Tobar et al. (Eds.): *Latest Advances in Electrical Engineering, and Electronics*, LNEE 933, pp. 129–140, 2022.
https://doi.org/10.1007/978-3-031-08942-8_10

real time [3]. For example, healthcare areas require high levels of lighting. At the same time, the impact that light has on the well-being of patients and staff, on the mental state and the recovery of people must be evaluated. Clarity is important for medical diagnosis and treatment. While visual performance for physicians must be balanced with the needs of patients. The current reality prevents the on-site evaluation of lighting levels in these areas, which is why wireless systems have become a necessity.

Control systems have variables that must be measured and monitored so that the control loop knows the actions to be performed in accordance with the data obtained from the sensors. In this case study we proposed the design and characterization of a wireless light meter based on IoT systems capable of transmitting its information to the cloud through the MQTT protocol, a lightweight, open protocol designed for low bandwidth and high latency networks, which guarantees a certain degree of security.

According to the International Commission on Illumination (CIE) a photometric instrument consists of a photometer head, a signal converter, an output device, and a power supply. The different parts can be built to a single device or split into separate housings [4]. Measurements of illuminance or luminance and their accuracy are influenced by various parameters, such as operational conditions, properties of light sources, as well as characteristics of the applied photometers as the photopic vision V(λ).

In order to meet this need, in this paper we present the design, implementation, and characterization of a low-cost wireless lighting meter, to obtain a prototype with the best possible performance. This paper is organized as follows. In Sect. 2, we describe the characteristics of the system detection, the device selection, the traceability system, and the test setup. In Sect. 3, we show the results of the design and characterization of the system, and the discussions of the data obtained. Finally, in Sect. 4 we present the conclusions and a brief over-view regarding possible future developments based on this work.

2 Development of the Illuminance Meter

2.1 Characteristics of the System Detection

Aiming to design a low-cost prototype, we define the detection system specifications. The operation spectral range should be according to the photopic range vision V(λ). This is the range in which a human eye is capable of distinguishing individual wavelengths. According to P. Boyce, the wavelength range from 380 nm to 780 nm is enough [5]. The illuminance range considered is 0 lx to 2000 lx, which is typical in photometric applications. The Signal to Noise Ratio (SNR) must be considered evaluated according to the spectral and spatial distribution of the radiation to be measured. Bandwidth has not been restricted because photometric measurements must be performed after a stabilization period of the light source, which can be up to 60 min depending on the type of light source [6]. From these features, we start the analysis.

Wireless communication is expanding at industrial, commercial, and residential levels. The need to implement wireless equipment in the field arises from the search to optimize automated monitoring systems, seeking greater efficiency and allowing the tracking of physical variables [7]. We proposed a prototype with a wireless communication module that allows the user to have two connection modes: direct connection through its access point mode and internet connection through the station mode, allowing to take data anywhere without affecting the measurements by lights or shadows introduced by the operator, in addition to real-time monitoring on any electronic device with internet access such as cell phones, pc, tablets, etc.

The accuracy of an optical detection system can be altered by the noise sources and the current they generated, typically denoted as current with a noise source suffix. The principal noise sources are the background radiation (I_{BG}), the shot noise (I_{SN}), the thermal noise (I_{TH}), and the electronic noise (I_E) [8]. The first one is due to the radiation emitted by the light sources and that is outside the visible range. This radiation gives rise to unwanted currents in the photodiode. The magnitude of this current depends on the responsivity of the sensor. The second one is related to the statistical fluctuation in the photocurrent and the dark current. The magnitude of the shot noise is expressed as

$$I_{SN} = \sqrt{2e(I_P + I_D)B}, \tag{1}$$

where e, I_P, I_D, and B are the electron charge, the bias current, the dark current, and the bandwidth. The thermal noise is due to the thermal generation of carriers. The magnitude of this generated noise current is

$$I_{TH} = \sqrt{\frac{4k_B TB}{R_{SH}}}, \tag{2}$$

where k_B, T, and R_{SH} are the Boltzmann constant, the temperature in degrees Kelvin, and the shunt resistance of the photodiode. Thermal noise is also present in all resistive elements, this is known as electronic noise (I_E). This is calculated using (2) and replacing R_{SH} with R_E. Finally, the total noise current (I_{TN}) generated is determined by

$$I_{TN} = \sqrt{I_{SN}^2 + I_{TH}^2 + I_E^2}. \tag{3}$$

Considering the features of the detection system and the noise sources, we chose the devices for a low-cost prototype.

2.2 Device Selection

We select the BH1750 photodetector because is a digital ambient light, based on a photodiode, a transimpedance amplifier, and a filter. Having the photodiode and amplifier integrated on a single chip reduces noise sources. The filter allows the passage of wavelengths between 400 nm to 700 nm, with which the sensor has spectral responsiveness close to the response of the human eye. According to CIE 023 "the contribution to the luminous responsivity due to the spectral responsivity at the borders of the visible wavelength range is small and the measurement uncertainties increase substantially" [4]. In addition, the sensor has a configurable high or low-resolution measurement system. The signal at the output of the amplifier is digitized by a 16-bit ADC and then sent by the I^2C bus interface to an ESP32 processor (Fig. 1a–b) [9].

ESP32 is a low-power system on a chip designed by Espressif Systems, that supports Wi-Fi, BT and BLE. The integrated chip has two CPU cores, each of them can be controlled individually. One of its main features is energy saving, for which it has a low-power processor that can be used in place of the CPU while performing tasks that do not require a lot of computing power. ESP32 integrates a wide set of peripherals, ranging from capacitive touch sensors, SD card interface, Ethernet, high-speed SPI, and I^2C. The sleep current is less than 5 μA, making it suitable for battery-powered and wear-able electronics applications. The module supports a data rate of up to 150 Mbps, and 20 dBm output power at the antenna to ensure the widest physical range [10]. These characteristics allowed us to make an application of electronic integration, low-power consumption, and wide range of connectivity, able to send the information to the network via Wi-Fi, that can be accessed with a mobile device (Fig. 1c). These characteristics allowed us to make an application of electronic integration, low-power consumption, and wide range of connectivity, able to send the information to the network via Wi-Fi, that can be accessed with a mobile device (Fig. 1c).

The interface was developed under the MVC pattern (Model – View – Controller) using HTML5. The application contains middleware to filter HTTP requests so that access to the administration module is limited to each registered user by their assigned role in the application. It also contains security for cross-site request forgery (CSRF) attacks to avoid unauthorized requests through registered users. The application consists of two main modules: administration and real-time measurements.

The prototype interface has two modules, the administration module, and the measurement module.

1. Administration Module:
    - Configuration of the access modes:
        - Access Point mode
        - Station Mode
    - Configuration of the MQTT broker information
    - User configuration
2. Measurement Module

Administration Module. It has three configurations' parameters: the configuration of the access mode, the configuration of the MQTT broker information and the configuration of users. A characteristic of this module is that its access is restricted by user password.

Access Mode Configuration. The prototype works with two access modes, the access point mode in which the prototype behaves as a wireless router allowing its direct connection with electronic devices (pc, tablets, or telephones) this mode is used when the operator is within the range of the prototype during the measurements. The station mode is used to connect to the internet through an access point allowing to monitor the measurements from anywhere in the world through MQTT brokers or platforms.

MQTT Broker Information Configuration. It allows to configure the necessary broker information to establish the communication through the MQTT protocol. MQTT Message Queue Telemetry Transport, is a publish/subscribe message transport protocol built on top of TCP to provide orderly and lossless bidirectional connections [11].

User Configuration. Allows adding users and passwords.

Measurement Module. This module displays the sensor readings in real time and is freely accessible.

The accuracy of any photodetector depends on the temperature. For this reason, we include in the device a temperature sensor DHT22 to correct the measure in case it is needed. The temperature dependency curve is in the datasheet of the photodetector [9].

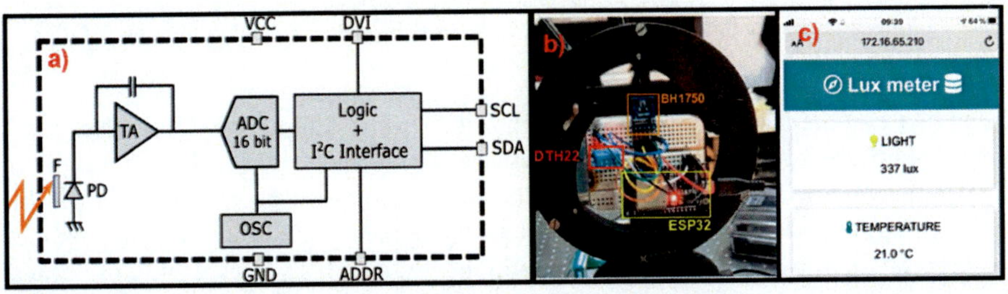

Fig. 1. (a) BH1750 schematic. Visible light filter (F), Photodiode (PD), Transimpedance amplifier (TA), AD converter (ADC) for obtainment digital 16-bit data, Logic + I^{2}C Interface. Ambient light calculation and I^{2}C BUS interface, include data register and measurement time register, Internal Oscillator (OSC). (b) Illuminance meter prototype mounted on the goniometer for the first test. Light sensor BH1750, temperature sensor DHT 22 and ESP32. (c) Screenshot of the mobile interface.

2.3 Test Setup and Traceability

To evaluate the performance of our photometer, we use the guide of CIE. Measurement of illuminance is influenced by various parameters, such as operational conditions, properties of light sources, as well as characteristics of the applied photometers [4]. To guarantee a repeatable test, we install the measuring system in an optical table, as it can see in Fig. 2. The device under test (DUT) is in a goniometer, the photodetector is the origin of the reference system. The optical axis x is the coordinate axis normal to the reference plane yz, from which distance measurements between lamp and photodetector are made. The optical center of the lamp and the photometer head are aligned to the optical axis. In the optical axis are located a diaphragm, polarizers, and filters, depending on the features being measured. The lamp is energized by a calibrated power supply (not shown) to avoid voltage fluctuations. The multimeter is connected to the cap of the bulb to check that the lamp is stabilized. As reference instrument for comparison, we use a calibrated lux meter. The data logger registers the ambient conditions to assurance that the test conditions have not changed significantly. We consider 20 °C as the optimal

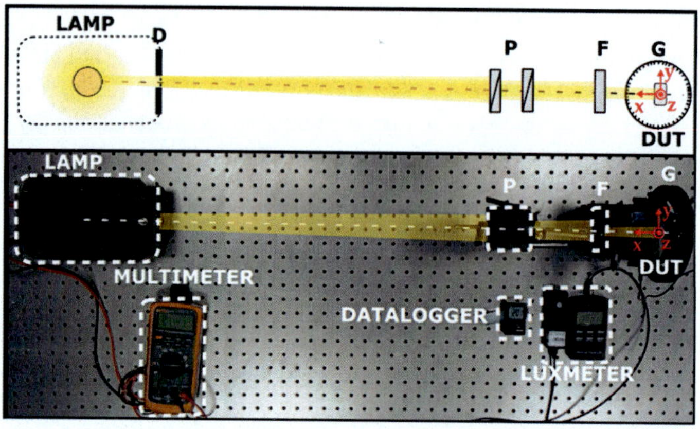

Fig. 2. Setup of measuring system. Lamp and DUT are aligned to optical axis. The multimeter and data logger assurance the condition of test. Polarizer (**P**), filter (**F**) and goniometer (**G**).

temperature of the test, because at this point the factor of temperature dependence is 1 [8]. Intending to avoid stray light, we cover the table with matte black fabric (not shown).

The uncertainty associated with the characterization of a photometer is a combination of the uncertainties arising from the measurement process and the uncertainties associated with the certified value of the reference standard. The traceability of our test is given by the instruments indicated below, which were calibrated by an accredited laboratory.

Reference Lux Meter. Mavolux 5032B is a high accurate instrument. It allows to measure small amounts of light, and it is provided with color correction. Its spectral response has matched 97% with the human eye V(λ). Cosine correction is also built in to ensure that oblique light incident, is also correctly assessed (98%). It has a linearity of 99% and an adjustment error of 0.8%. This results in a maximum error of 8%. The lux meter has been calibrated by DAkkS, which is the national accreditation authority of the Federal Republic of Germany.

Intensity Pattern. Incandescent lamp, kept in a controlled environment to preserve its useful life, is used only for verification of working standards and to provide traceability. When it is energized a 158.1 V and 1.1097 A, it radiates a luminous intensity of 267.4 cd, with a Reference Color Temperature of 2856°K. This standard is traceable to the National Instrument of Metrology (NIM) and we use as CIE A illuminant.

Stabilized Voltage Source. It is calibrated by Centro de Metrología del Ejército Ecuatoriano. To give continuity to the SI traceability, this equipment is verified twice a year with a calibrated electrical analyzer.

Multimeter. The equipment is traceable to the International System of Units (SI) and is calibrated every year. It has been calibrated in AC/DC voltage, AC/DC current, and resistance.

Thermo Hygrometer. To monitor environmental conditions during measurements. Temperature accuracy ±0.5 °C (−20 to +70 °C) and Humidity accuracy ±3% RH (2% RH to 98% RH) ±1 digit +0.03% RH/K. To ensure traceability to the SI, the measuring instrument is calibrated for temperature and humidity every year.

Spectrophotometer. Spectrum analyzer, LED color meter, and flicker meter for lighting systems. It is integrated with spectral technology and optimizes the LUX measurement range. It has a measurement range of 5 to 70000 lx, a wavelength range of 380 to 780 nm, and a wavelength data increment of 1 nm. This instrument is not calibrated and is used for informational purposes only. From what is described in this section, we carry out the characterization of the photometer and evaluate the results.

2.4 Test Lamps

To evaluate the performance of the illuminance meter, we tested it with different spectra. We used four lamps of different technology and Correlated Color Temperature (TCC) (Fig. 3). An incandescent lamp 60 W 2850°K, a compact fluorescent lamp (CFL) 23 W 2691°K, a neutral white LED lamp 8 W 3910°K, and a warn white LED lamp 9 W 2705°K. All the lamps were previously seasoned according to IESNA Guide to Lamps Seasoning to guarantee the stabilization of photometric parameters [12] and energized at 120 V for the test. Figure 3 shows the specters obtained from the described lamps using the spectrometer.

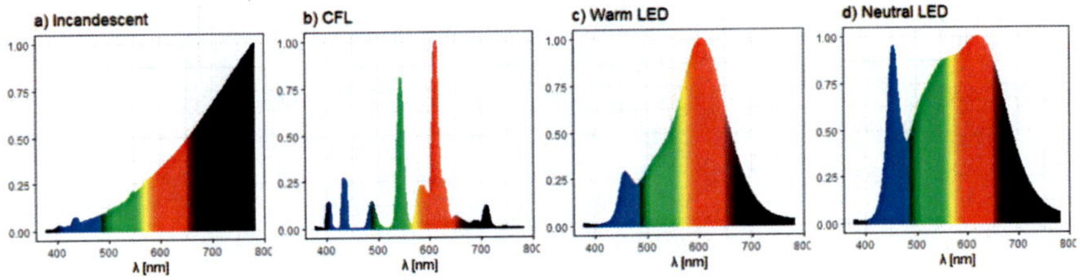

Fig. 3. Real specters from lamps used.

3 Test Results and Discussions

The prototype administration module, which allows the configuration of the two connection modes, station, and access point, is tested with satisfactory results. The correct functioning of the embedded interface is also verified by means of HTTP requests with the use of different browsers in different electronic devices such as PCs, telephones, and tablets, allowing to verify the compatibility of the interface in these devices. The MQTT server is configured in the prototype (broker address, the topic format, API key) to test the data transmission to the cloud; the prototype's real-time measured data can be observed with the help of the ThingSpeak for IoT Projects by Mathworks platform.

Its wireless connection to IoT platforms via MQTT allows real-time data to be obtained that can be used for other applications in more complex systems such as smart lighting systems.

For a photometric measurement using a photometer whose spectral responsivity $s_{rel}(\lambda)$ differs in certain spectral ranges from the photopic vision $V(\lambda)$, incorrect measurement results can be obtained. The $s_{rel}(\lambda)$ of the photometer was scanned from its datasheet [8], and the $V(\lambda)$ corresponds to (Fig. 4) [4]. The difference between $V(\lambda)$ and $s_{rel}(\lambda)$ causes the aforementioned error.

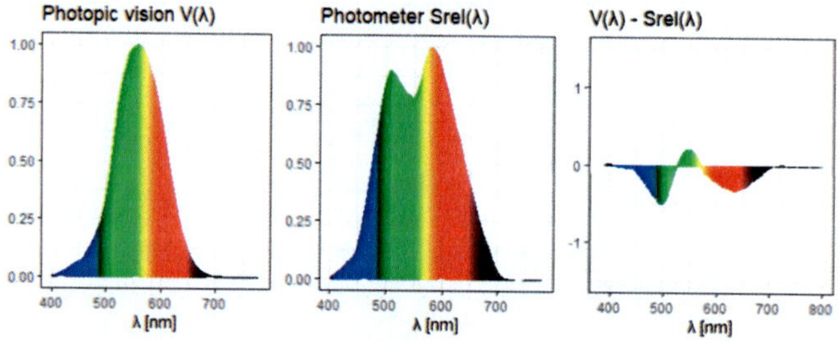

Fig. 4. Photopic vision $V(\lambda)$, and the responsivity of the photometer $s_{rel}(\lambda)$. The difference between them causes an error in the measurements.

Photometers are normally calibrated with a CIE Source A lamp. However, according to CIE 23 "when using the spectrally integrated responsivity function, such differences may compensate each other to some extent when comparing two spectral distributions, e.g. light source Z and CIE Standard Illuminant A". In order to correct the measurements, we introduce an adjustment factor in the measurement system. We calculate the relative luminous responsivity $a^*(s_Z(\lambda))$, that is the ratio of the luminous responsivity s_Z when the detector is illuminated with light source Z to the luminous responsivity s_A when it is illuminated with CIE Standard Illuminant A, and is calculated as

$$a^*(S_z(\lambda)) = \frac{s_Z}{s_A} = \frac{\int_{\lambda_{max}}^{\lambda_{min}} S_Z(\lambda)S_{ref}(\lambda)d\lambda}{\int_{360\ nm}^{830\ nm} S_Z(\lambda)V(\lambda)d\lambda} / \frac{\int_{\lambda_{max}}^{\lambda_{min}} S_A(\lambda)S_{ref}(\lambda)d\lambda}{\int_{360\ nm}^{830\ nm} S_A(\lambda)V(\lambda)d\lambda}, \quad (4)$$

where λmin and λmax should cover the entire range where $s_Z(\lambda)$ $s_{rel}(\lambda)$ has non-zero values. s_A and s_Z correspond to the spectrums of a CIE Standard Illuminant A, and the neutral white LED lamp. We used this method because it was the way we got the least error.

Aiming to characterize the photometer, we perform measurements using the system in Fig. 2, and the test lamps described previously. When comparing the photodetector measurements with those of the reference's lux meter, we detect a deviation as can see in Fig. 5a. It shows the distribution of errors for each type of lamp is different. With the adjustment factor, the lowest error is obtained for solid-state lamps (LED), which corresponds to approximately ±2%. This is a great advantage since solid-state light is the most common lighting technology today. On the other hand, with incandescent and CFL lamps, the error is greater. This may be since we use a dimmer to regulate the luminous flux of these lamps. In the case of the incandescent lamp, this is due to the response capacity of the photodetector is zero starting at 700 nm, where the lamp has its highest emission. This makes the measurement smaller and therefore the mean value of the error is 10% approximately. The outliers obtained are few, are marked with a red asterisk, and can be attributed to stray lights during the experiment. Despite the errors obtained, the linearity of the photometer is high (Fig. 5b). As can be seen in the figure, the black points correspond to the measured values and the blue line corresponds to the fit curve. The correlation coefficient obtained is 0.99.

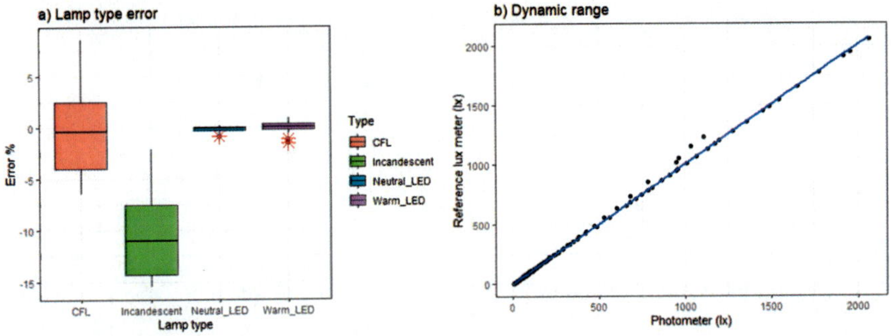

Fig. 5. a) Distribution of the error obtained by type of lamp using $a^*(s_Z(\lambda))$. b) Dynamic range and detector linearity.

From the data obtained, we established 2 lx as the limit of detection (LOD), because it was not possible to achieve lower stable values. From this result, we can deduce that the total noise is lesser than 2 lx. On the other hand, the measurement is zero in complete darkness. This leads us to think that, under the test ambient conditions, the darkness photocurrent and the thermal noise are lesser than the resolution of the ADC. Thus, the SNR is large enough to obtain stable measurements. To cover the proposed dynamic range, we use the high resolution (0.1 lx) in measurements lesser than 100 lx and the low resolution (1 lx) in measurements greater. The range change is programmed to be done automatically.

Another important factor to determine in the photodetector is the spectral and spatial distribution. From the BH1750 scheme (Fig. 1a) and the responsivity curve (Fig. 4b), we can say that the F filter prevents ultraviolet (UV) and infrared (IR) radiation from reaching the photodetector. For this reason, we consider the background noise to be zero. This was checked using UV and IR filters between the lamps and the photodetector. On the other hand, checking the spatial distribution was necessary to evaluate the cosine response. This determines the accuracy of the measurement for light that arrives at angles other than the normal to the photodetector. Using a goniometer, we rotate the photometer and measure the illuminance every 10°. We carry out this procedure for the horizontal orientation (*xy* plane) and the vertical orientation (*xz* plane). The results are plotted in the Fig. 6, where red, green, and grey lines represent the vertical, horizontal, and ideal response respectively, and the 0° corresponds to the optical axis. As it can see, the vertical and horizontal responses are similar, but if they are com-pared with the ideal response there is a deviation. Such deviation could lead to errors, this is important when measuring lighting installations such as office lighting or street lighting where the light comes from different angles.

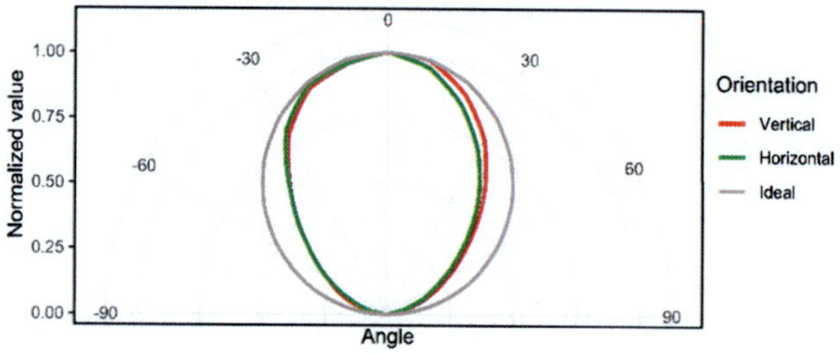

Fig. 6. Cosine response for the photometer.

Specular reflections and certain luminaries may cause the light to be polarized. To measure the polarization dependence, we placed two sheet-polarizers back-to-back with their axes parallel in front of the light source and rotate one of them. The illuminance obtained by rotating the polarizer agrees with Malus's law. The maximum illuminance measured is less than the illuminance without polarization in approximately 60%, and the spectrum of the lamp was not affected. However, at the lower value of illuminance, we could observe that the wavelengths in IR and UV are the least attenuated as seen in Fig. 7. The polarization dependence index is calculated as

$$f(\varepsilon, \varphi) = \frac{E_{max}(\varepsilon, \varphi) - E_{min}(\varepsilon, \varphi)}{E_{max}(\varepsilon, \varphi) + E_{min}(\varepsilon, \varphi)} = 0.97, \quad (5)$$

where ε is the angle of incidence measured from the optical axis (0° for our case) and φ is the azimuth angle. Under the conditions of our experiment, we did not find any other dependencies that we can mention.

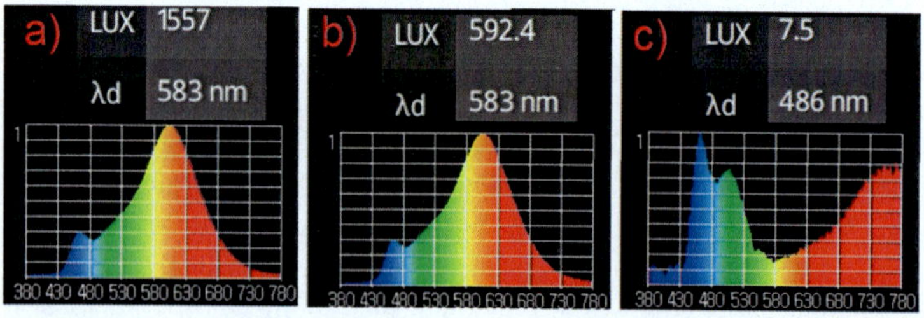

Fig. 7. a) Emission spectrum without a polarizer. b) Spectrum with polarizer at the highest illuminance angle. c) Spectrum with polarizer at the angle of least illuminance.

If the photometer is exposed to constant illumination over long periods, it is necessary to determine the temporary change in the responsivity, this factor is called fatigue index. Under the test conditions described above, we exposed the photometer and reference lux meter to a constant illuminance for 3 h. The illuminance level was 2650 lx, which is a value over the designed range, and the period time was established considering 3 times the time that the test of a lamp lasts. The fatigue index is calculated as

$$f(t) = \frac{E(t = 3h)}{E(t_0)} - 1 = 0, \quad (6)$$

where E(t) is the illuminance at time t, and to is the reference time, for our case it is 5 s. As can see, there is no difference between the first and the last measure.

The ambient temperature influences the relative spectral responsivity of the photometer. For our case, we could not achieve controlled variations in temperature that are large enough to show significant changes in the illuminance. However, we implement the algorithm to compensate for the measuring using the DHT22 sensor. Theoretically, when the temperature increases at 80 °C the illuminance increments 1.05 times. On the other hand, when the temperature decreases at −40 °C the illuminance reduces 0.95 times. The ideal temperature of the photometer is 20 °C. Finally, under the explained conditions, we consider that the implemented wireless illuminance meter is capable to use for many applications and experiences.

4 Conclusions

The characterization of the photodetector system allows the design of an illuminance meter guaranteeing the accuracy of the measurements, in addition the ESP32 through its Wi-Fi module gives it the ability to connect to the internet. In this way the data can be accessed in real time from anywhere in the world to be used for different purposes and applications.

The accuracy of the optical detection system depends on the noise sources and the current generated in the photodetector. The main noise sources are identified and using different spectra it is found that the background noise is reduced by the photodetector filter, which is demonstrated by the responsiveness of the photometer. The shot noise is low because the bias current of the photodiode and the bandwidth of the measurement system are low. Additionally, the dark current and the thermal noise are lesser than the resolution of the ADC in the high-resolution mode, so they do not affect the quality of the measurement.

The characterization of the photodetector system allows the design of an illuminance meter guaranteeing the accuracy of the measurements, in addition the ESP32 through its Wi-Fi module gives it the ability to connect to the internet. In this way the data can be accessed in real time from anywhere in the world to be used for different purposes and applications. The main noise sources are identified and using different spectra it is found that the background noise is reduced by the photodetector filter, which is demonstrated by the responsiveness of the photometer.

The performance of the photodetector is limited by its response curve, as it does not perfectly match the photopic viewing curve. Measurements are adjusted using the relative luminous responsivity a*(sz(λ)) suggested by CIE 23 to compensate and reduce the error. It is known that photometers are normally calibrated with a CIE Source A lamp; however, using a* yields the lowest error for CFL and LED lamps which are the most commonly used today. The spatial distribution response of the photodetector does not match the ideal cosine response, which could lead to errors. This is important when measuring lighting installations, such as office lighting or street lighting, where the light comes from different angles. Caution with this factor is recommended. Furthermore, the experimental results show that the designed illuminance meter is not affected by fatigue or polarized light. Furthermore, the experimental results show that the designed illuminometer is not affected by fatigue, polarized light and small temperature variations.

References

1. Wu, F., Wu, T., Yuce, M.R.: Design and implementation of a wearable sensor network system for iot-connected safety and health applications. In: IEEE 5th World Forum Internet Things, WF-IoT 2019 – Conf Proc, pp. 87–90 (2019). https://doi.org/10.1109/WF-IoT.2019.8767280
2. Obuchi, Y., et al.: Measurement and evaluation of comfort levels of apartments using IoT sensors. In: 2018 IEEE Int Conf Consum Electron ICCE 2018, Jan, pp. 1–6 (2018). https://doi.org/10.1109/ICCE.2018.8326169
3. Ahmad, M.D., Noor, S.Z.M., Rahman, N.F.A., Haris, F.A.: Lux meter integrated with internet of things (IoT) and data storage (LMX20). In: ICPEA 2021 – 2021 IEEE Int Conf Power Eng Appl, pp. 138–142 (2021). https://doi.org/10.1109/ICPEA51500.2021.9417762
4. ISO/CIE 19476:2014.: Characterization of the performance of illuminance meters and luminance meters (2014). http://www.cvrl.org/database/data/lum/linCIE2008v2e_1.csv
5. Braun, M., et al., Human Factors in Lighting BT – Ergonomics and Health Aspects of Work with Computers. In: Karsh, B.-T. (ed.) Springer Berlin Heidelberg, Berlin, Heidelberg, pp. 223–230 (2009)
6. De La Bastida, R., Araguillin, R., Sotomayor, N., Chasi, C.: Implementation of a real-time monitoring and analysis system for luminous flux test in integrating sphere. In: Narváez, F.R., Vallejo, D.F., Morillo, P.A., Proaño, J.R. (eds.) SmartTech-IC 2019. CCIS, vol. 1154, pp. 16–28. Springer, Cham (2020). https://doi.org/10.1007/978-3-030-46785-2_2
7. Tsiatsis, V., et al.: Technology Fundamentals (2019)
8. Rogalski, A., Bielecki, Z. Detection of optical radiation. Bull. Polish Acad. Sci. Tech. Sci. **52**(1) (2006)
9. ROHM semiconductor: Digital 16bit Serial Output Type Ambient Light Sensor IC 21. (2011)
10. Espressif: https://www.espressif.com/
11. Toapanta, R., Chafla, J., Toapanta, A.: Physical variables monitoring to contribute to landslide mitigation with IoT-based systems BT. In: In: Botto-Tobar M.S., Gómez, O., Rosero Miranda, R., Díaz Cadena, A. (eds.) Advances in Emerging Trends and Technologies. Springer International Publishing, Cham, pp. 58–71 (2021)
12. IES LM-54-12: IES guide to lamp seasoning. J. Illum. Eng. Soc. 7, 144–146. (1978). https://doi.org/10.1080/00994480.1978.10747833

Decoupled Distributed State Estimator with Reduced Number of Measurements for Power System Applications

Javier Almeida[1(✉)], Silvana Gamboa[1,2], and Jackeline Abad[1]

[1] Escuela Politécnica Nacional, Quito, Ecuador
{darwin.almeida,silvana.gamboa,jackeline.abad}@epn.edu.ec
[2] GIECAR, Escuela Politécnica Nacional, Quito, Ecuador

Abstract. This paper presents the formulation of a decoupled distributed state estimator using a reduced number of measurements. This proposal aims to solve a distributed state estimation in which the estimation of each area of the system will be performed locally. The measurements made by the PMUs installed in different nodes defined as references are included in the solution of the state estimator. The estimator is based on the weighted least squares (WLS) method. Furthermore, we intend to use a reduced number of measurements during the estimation, so we use a minimum spanning tree to provide the minimum amount of measurements that guarantee the observability of each area to be estimated. Consequently, the proposed state estimator uses only the transferred power and phasor voltages aggregating the phase angles of the voltages in each area of the decomposed system. In order to validate the performance of the implemented distributed estimator, the states obtained for the IEEE-39 bus system case study are compared with the states of an implementation of the system in the PowerFactory electrical analysis tool. The results show an adequate performance of the estimator. Consequently, the feasibility of a distributed estimation that can be performed in a decoupled way and with a reduced number of measurements is verified.

Keywords: Distributed state estimation · Decoupled estimation · Minimum spanning tree · Phasor measurement unit · Weighted least squares

1 Introduction

Nowadays, the power system includes different types of power generation located close to consumption centers, hence its dynamical components are now distributed throughout the power system [1–3]. Under such conditions, control strategies must also change and adapt to the characteristics of a distributed

The authors acknowledge the support from Escuela Politécnica Nacional research grant PIS-19-01.

M. Botto-Tobar et al. (Eds.): *Latest Advances in Electrical Engineering, and Electronics*, LNEE 933, pp. 141–153, 2022.
https://doi.org/10.1007/978-3-031-08942-8_11

power system. As a result, several distributed control techniques have been proposed in recent years [2,4]. However, the state estimation in power systems is still executed centrally as a component of energy management systems in a control center.

Even if distributed controllers could use the results of the estimator executed in the control center, their centralized characteristic must be taken into account. In other words, although the controller would be distributed, the state monitoring in a control area will still depend on a centralized application, which has its disadvantages such as time delays. Therefore, a state estimation that is also executed in a distributed manner is necessary to achieve a fully distributed control system. As a result, a distributed state estimation allows to reduce estimation times and achieve the real-time control required by the current electrical system. Different methods for distributed state estimation have been proposed. For instance, [5–7] develop a two-step estimation: firstly, the power system is split in smaller estimation areas, where local estimation is performed; then, the estimated values from previous step are integrated by different methods. Additionally, [8] uses an equal-to-equal decomposition method of Karush-Kuhn-Tucker (KKT) condition to formulate the fully distributed fast decoupled state estimation (FD2SE) algorithm. While [9] uses the extended Kalman filter (EKF) and weighted least squares (WLS) considering weekly variations of bus power injection. Although distributed estimation is accomplished, high processing times or complex developments are required. In this regard, this paper aims to provide a solution to the problem by formulating a one-stage distributed state estimator, which uses a minimum number of measurements and does not require information exchange with neighboring local estimators.

Our proposed method considers synchronized phasor measurements performed by PMUs, which are located at nodes known as *references* of each area to be estimated including the slack node of the system, to provide a common angular reference for all estimation areas. This condition allows each area to estimate its states, which reflects the state of the entire system, without consulting the neighboring estimators or exchanging data with them. So, unlike aforementioned proposed methods, our proposal does not make use of measurements at the boundary buses between areas or data exchange with neighboring areas. Further, the division of the system areas can be arbitrary considering the existence of at least one PMU. In addition, we consider a minimum number of measurements by implementing a minimum spanning tree criterion to reduce measurements during processing and ensure the observability of the area. In order to validate our proposal, the estimator is implemented in a numerical analysis software and validated on the IEEE-39 bus system. The results demonstrate the feasibility of the proposal and thanks to its characteristics, this proposal could be a fast way to provide current state variables to distributed control algorithms implemented in power systems.

This work is organized as follows. Section 2 presents a review of state estimation in power system. Then, the proposed estimator is formulated in Sect. 3. In Sect. 4, procedure for estimator implementation is detailed. In Sect. 5, tests and their results are presented. Finally, the conclusions are drawn in Sect. 6.

2 State Estimation in Power Systems

2.1 General Features

The state estimator (SE) establishes the state of the electrical system by obtaining the voltage phasors in all the nodes of the system using the information coming from the SCADA system [10]. It includes: sufficient measurements of different electrical variables so that the algorithm in the control center is able to obtain the solution of the system, and the information and topology of the electrical network [10]. The state estimation can be centralized or distributed. However, if the architecture is centralized, there are disadvantages because of the information handling and execution time, such problems are solved when performing a distributed estimation.

Nowadays, measurements of electrical system variables are supported by phasor measurement units (PMUs), which are used to perform synchronized measurements of the voltage phasor on buses and the currents adjacent to the branches of the bus. The measurements generated by the PMUs are synchronized through a GPS (global positioning system) signal. The voltage and current phasors obtained have the same time reference so they can be used to know the real state of the system at a given time instant.

2.2 WLS Formulation for State Estimation

The WLS method is widely used for state estimation in power systems because the precision of each type of measurement is different. Since WLS can assign a weight according to the type of measurement through a matrix of weights, it improves the precision of the estimate.

We begin to formulate the proposed distributed estimator using the above-mentioned method. First, the measurements of the electrical variables are expressed in terms of the system state variables by Eq. (1).

$$z = \begin{bmatrix} z_1 \\ z_2 \\ \vdots \\ z_n \end{bmatrix} = \begin{bmatrix} h_1(x_1, x_2, s, x_n) \\ h_2(x_1, x_2, s, x_n) \\ \vdots \\ h_n(x_1, x_2, s, x_n) \end{bmatrix} + \begin{bmatrix} e_1 \\ e_2 \\ \vdots \\ e_n \end{bmatrix} = h(x) + e, \tag{1}$$

where z represents the measured value, $h(x)$ represents the measurement function vector containing linear and nonlinear equations relating the system states to the measured values, x represents the system state vector including voltage and angle of all nodes except the angle at the slack node and e represents the error. x_i and e_i represent the ith state and the error in the ith measurement respectively.

According to the WLS method, the objective function (Eq. (2)) is the sum of the squares of the measurement errors [11].

$$J(x) = [z - h(x)]^T R^{-1} [z - h(x)], \tag{2}$$

where the weight matrix R^{-1} is composed of the reciprocals of the error variances of the corresponding measurements as is shown in Eq. (3),

$$R^{-1} = \begin{bmatrix} \frac{1}{\sigma_1^2} & 0 & 0 & 0 \\ 0 & \frac{1}{\sigma_2^2} & 0 & 0 \\ 0 & 0 & \ddots & 0 \\ 0 & 0 & 0 & \frac{1}{\sigma_n^2} \end{bmatrix}. \tag{3}$$

Next, we establish a local minimum to obtain an optimal condition by equating the first derivative to zero. As a result, Eq. (4) is obtained.

$$g(x) = \frac{\partial J(x)}{\partial x} = -H^T R^{-1}[z - h(x)] = 0, \tag{4}$$

where H is the Jacobian matrix of the nonlinear functions in Eq. (5).

$$H = \frac{\partial h(x)}{\partial x} = \begin{bmatrix} \frac{\partial h_1}{\partial x_1} & \frac{\partial h_1}{\partial x_2} & s & \frac{\partial h_1}{\partial x_n} \\ \frac{\partial h_2}{\partial x_1} & \frac{\partial h_2}{\partial x_2} & s & \frac{\partial h_2}{\partial x_n} \\ \vdots & \vdots & \ddots & \vdots \\ \frac{\partial h_m}{\partial x_1} & \frac{\partial h_m}{\partial x_2} & s & \frac{\partial h_m}{\partial x_n} \end{bmatrix}. \tag{5}$$

Next, we proceed to linearize the function $g(x)$ around a state vector x^k by Taylor series. As a result, Eq. (6) is obtained.

$$g(x^k) + \partial g(x^k)/\partial x^k \triangle x = 0, \tag{6}$$

where its terms are:

$$g(x^k) = -H^T R^{-1}[z - h(x^k)] \tag{7}$$

and

$$\partial g(x^k)/\partial x^k = -H^T R^{-1} H = -G \tag{8}$$

In Eq. (6), we solve for the differential and consider Eqs. (7) y (8).

$$\triangle x = G^{-1} H^T R^{-1}[z - h(x^k)] \tag{9}$$

The differential is also defined by the subtraction of the posterior state and the anterior state in Eq. (10).

$$\Delta x = x^{k+1} - x^k \tag{10}$$

Finally, to obtain the next state of the state vector we substitute Eq. (10) in Eq. (9). The iterative process to be performed in each iteration will be the one defined in Eq. (11).

$$x^{k+1} = G(x^k)^{-1} H(x^k)^T R^{-1}[z - h(x^k)] + x^k \tag{11}$$

The iterative process consists of finding the state that minimized the distance from the obtained measurements to the estimated measurements that is the minimum value of Δx.

3 Decoupled Distributed State Estimator Formulation

3.1 Measurement Functions

As shown in Eq. (11), the estimation depends on the measurement functions $h(x)$ and their Jacobian $H(x)$, defined according to the type of measurements. For this reason, equations that characterize every type of measurements must be obtained. In this proposal, unlike the classical formulation of the state estimator, the measurements of the PMUs (magnitude and phase angle voltage) will be considered because they are synchronized in time with PMUs in other areas, which allows to reference the distributed state estimation to a common reference (slack node). The measurement functions for the PMUs along with the power flow measurements are shown below.

Node Voltage. Eq. (12) shows the value of the voltage in magnitude at a node k measured by the PMU.

$$h_1(x) = V_k \tag{12}$$

Node Angle. Eq. (13) shows the value of the angle at a node k measured by the PMU.

$$h_2(x) = \Theta_k \tag{13}$$

Transmission Line Power Flow. We apply Kirchhoff's laws on the circuit that represents the transmission line of medium length (Fig. 1) to obtain the equations describing the flow of active and reactive power through them. The active power and reactive power flow are shown in Eq. (14) and Eq. (15) respectively.

$$h_3(x) = P_{km} = -V_k^2 G_{km} + V_k V_m (G_{km} \cos\Theta_{km} + B_{km} \sin\Theta_{km}) \tag{14}$$

$$h_4(x) = Q_{km} = V_k^2 (B_{km} - bs_{km}) + V_k V_m (G_{km} \sin\Theta_{km} - B_{km} \cos\Theta_{km}) \tag{15}$$

Transformer Power Flow. The model of the transformer is shown in Fig. 2. After using Kirchhoff's laws, we obtain the active power flow km in Eq. (16), reactive km in Eq. (17), active mk in Eq. (18) and reactive mk in Eq. (19).

$$h_5(x) = PT_{km} = -V_k^2 G_{km}/a_{km} + V_k V_m (G_{km} \cos\Theta_{km} + B_{km} \sin\Theta_{km}) \tag{16}$$

$$h_6(x) = QT_{km} = V_k^2 B_{km}/a_{km} + V_k V_m (G_{km} \sin\Theta_{km} - B_{km} \cos\Theta_{mk}) \tag{17}$$

$$h_7(x) = PT_{mk} = -V_m^2 G_{mk} a_{km} + V_m V_k (G_{mk} \cos\Theta_{mk} + B_{mk} \sin\Theta_{mk}) \tag{18}$$

$$h_8(x) = QT_{mk} = V_m^2 B_{mk} a_{km} + V_m V_k (G_{mk} \sin\Theta_{mk} - B_{mk} \cos\Theta_{mk}) \tag{19}$$

The vector $h(x)$ (Eq. (20)) depends on the order of measurement vector (z) of the SCADA.

$$h(x) = \left[h_1(x)\ h_2(x)\ h_3(x)\ h_4(x)\ h_5(x)\ h_6(x)\ h_7(x)\ h_8(x)\right]^T \tag{20}$$

3.2 Jacobian Elements

The elements that are part of the structure of the Jacobian consist of the derivatives of the measurements with respect to the state variables (angles and voltages, except for the ones related to the reference nodes). The structure that defines the Jacobian $H(x)$ is shown in Eq. (21).

$$H = \begin{bmatrix} H_{11} & H_{12} \\ H_{21} & H_{22} \\ H_{31} & H_{32} \\ H_{41} & H_{42} \\ H_{51} & H_{52} \\ H_{61} & H_{62} \\ H_{71} & H_{72} \\ H_{81} & H_{82} \end{bmatrix} \tag{21}$$

Jacobian Voltage Elements. The contribution to the Jacobian is given by the partial derivative of the voltage magnitude with respect to the angles (Eq. (22)) and with respect to voltages (Eq. (23)).

$$H_{11} = \partial \left\{V_k\right\} / \partial \Theta_k = 0 \tag{22}$$

$$H_{12} = \partial \left\{V_k\right\} / \partial V_k = 1 \tag{23}$$

Jacobian Angle Elements. The contribution to the Jacobian is given by the partial derivative of the node angle with respect to the angles (Eq. (24)) and with respect to the voltages (Eq. (25)).

$$H_{21} = \partial \left\{\Theta_k\right\} / \partial \Theta_k = 1 \tag{24}$$

$$H_{22} = \partial \left\{\Theta_k\right\} / \partial V_k = 0 \tag{25}$$

Jacobian Elements of Transmission Line Power Flow. The contribution to the Jacobian is composed of H_{31} in Eqs. (26) and (27), H_{32} in Eqs. (28) and (29) H_{41} in Eqs. (30) and (31) and H_{42} in (32) and (33).

$$\partial P_{km} / \partial \Theta_m = V_k V_m (G_{km} \sin \Theta_{km} - B_{km} \cos \Theta_{km}) \tag{26}$$

$$\partial P_{km} / \partial \Theta_k = V_k V_m (-G_{km} \sin \Theta_{km} + B_{km} \cos \Theta_{km}) \tag{27}$$

$$\partial P_{km} / \partial V_m = V_k (G_{km} \cos \Theta_{km} + B_{km} \sin \Theta_{km}) \tag{28}$$

$$\partial P_{km} / \partial V_k = -2 V_k G_{km} + V_m (G_{km} \cos \Theta_{km} + B_{km} \sin \Theta_{km}) \tag{29}$$

$$\partial Q_{km} / \partial \Theta_m = -V_k V_m (G_{km} \cos \Theta_{km} + B_{km} \sin \Theta_{km}) \tag{30}$$

$$\partial Q_{km} / \partial \Theta_k = V_k V_m (G_{km} \cos \Theta_{km} + B_{km} \sin \Theta_{km}) \tag{31}$$

$$\partial Q_{km} / \partial V_m = V_k (G_{km} \sin \Theta_{km} - B_{km} \cos \Theta_{km}) \tag{32}$$

$$\partial Q_{km} / \partial V_k = 2 V_k (B_{km} - bs_{km}) + V_m (G_{km} \sin \Theta_{km} - B_{km} \cos \Theta_{km}) \tag{33}$$

Jacobian Elements of Transformer Power Flow. The contribution to the Jacobian is composed of H_{51} in Eqs. (34) and (35), H_{52} in Eqs. (36) and (37), H_{61} in Eqs. (38) and (39), H_{62} in Eqs. (40) and (41), H_{71} in Eqs. (42) and (43), H_{72} in Eqs. (44) and (45), H_{81} in Eqs. (46) and (47) and H_{82} in Eqs. (48) and (49).

$$\partial PT_{km}/\partial \Theta_m = V_k V_m (G_{km} \sin \Theta_{km} - B_{km} \cos \Theta_{km}) \tag{34}$$

$$\partial PT_{km}/\partial \Theta_k = V_k V_m (-G_{km} \sin \Theta_{km} + B_{km} \cos \Theta_{km}) \tag{35}$$

$$\partial PT_{km}/\partial V_m = V_k (G_{km} \cos \Theta_{km} + B_{km} \sin \Theta_{km}) \tag{36}$$

$$\partial PT_{km}/\partial V_k = -2V_k G_{km}/a_{km} + V_m (G_{km} \cos \Theta_{km} + B_{km} \sin \Theta_{km}) \tag{37}$$

$$\partial QT_{km}/\partial \Theta_m = -V_k V_m (G_{km} \cos \Theta_{km} + B_{km} \sin \Theta_{km}) \tag{38}$$

$$\partial QT_{km}/\partial \Theta_k = V_k V_m (G_{km} \cos \Theta_{km} + B_{km} \sin \Theta_{km}) \tag{39}$$

$$\partial QT_{km}/\partial V_m = V_k (G_{km} \sin \Theta_{km} - B_{km} \cos \Theta_{km}) \tag{40}$$

$$\partial QT_{km}/\partial V_k = 2V_k B_{km}/a_{km} + V_m (G_{km} \sin \Theta_{km} - B_{km} \cos \Theta_{km}) \tag{41}$$

$$\partial PT_{mk}/\partial \Theta_k = V_m V_k (G_{mk} \sin \Theta_{mk} - B_{mk} \cos \Theta_{mk}) \tag{42}$$

$$\partial PT_{mk}/\partial \Theta_m = V_m V_k (-G_{mk} \sin \Theta_{mk} + B_{mk} \cos \Theta_{mk}) \tag{43}$$

$$\partial PT_{mk}/\partial V_k = V_m (G_{mk} \cos \Theta_{mk} + B_{mk} \sin \Theta_{mk}) \tag{44}$$

$$\partial PT_{mk}/\partial V_m = -2V_m G_{mk} a_{km} + V_k (G_{mk} \cos \Theta_{mk} + B_{mk} \sin \Theta_{mk}) \tag{45}$$

$$\partial QT_{mk}/\partial \Theta_k = -V_m V_k (G_{mk} \cos \Theta_{mk} + B_{mk} \sin \Theta_{mk}) \tag{46}$$

$$\partial QT_{mk}/\partial \Theta_m = V_m V_k (G_{mk} \cos \Theta_{mk} + B_{mk} \sin \Theta_{mk}) \tag{47}$$

$$\partial QT_{mk}/\partial V_k = V_m (G_{mk} \sin \Theta_{mk} - B_{mk} \cos \Theta_{mk}) \tag{48}$$

$$\partial QT_{mk}/\partial V_m = 2V_m B_{mk} a_{km} + V_k (G_{mk} \sin \Theta_{mk} - B_{mk} \cos \Theta_{mk}) \tag{49}$$

The values of $h(x)$ and $H(x)$ depend on the conductance, G, and susceptance, B, values from the area admittance matrix (Y_{bus}). The area admittance matrix is formed by transmission line admitance (Eqs. (50)) and transformer admitance ((51)) matrices, whose models are shown in Fig. 1 and 2. The value of a represents the value of the transformer ratio (equivalent to 1 in the estimator when it is 0 in the SEP) and bs represents half of the admittance of the shunt branch (imaginary part).

$$Y_{line} = \begin{pmatrix} Y_{km} + bs_{km} & -Y_{km} \\ -Y_{mk} & Y_{mk} + bs_{mk} \end{pmatrix} \tag{50}$$

$$Y_{trafo} = \begin{pmatrix} Y_{km}/a^2 & -Y_{km}/a \\ -Y_{mk}/a & Y_{mk} \end{pmatrix} \tag{51}$$

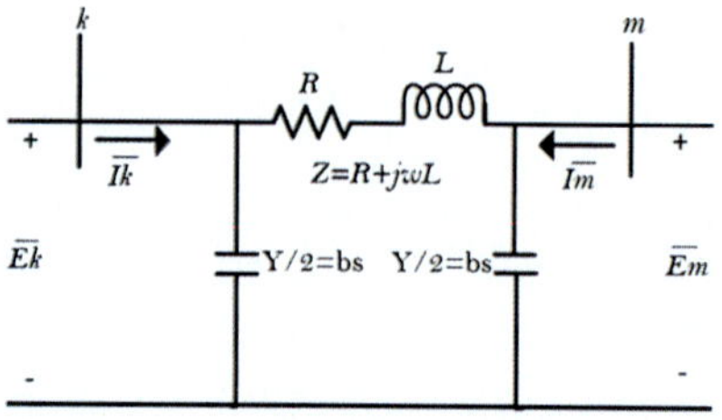

Fig. 1. Transmission line model.

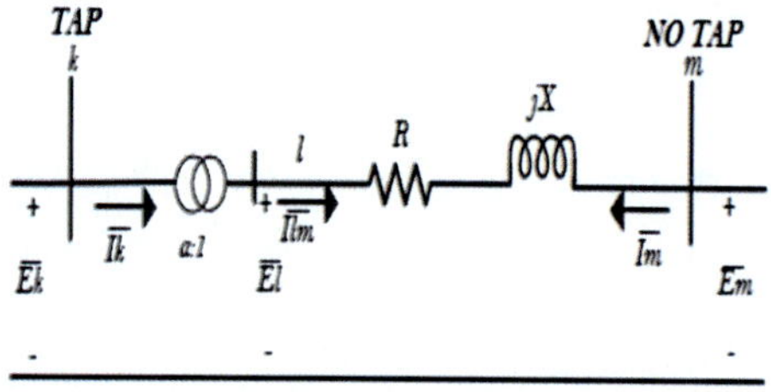

Fig. 2. Transformer model.

3.3 Minimum Spanning Tree for Reduction of Measurements

The estimation requires that the area is observable. To achieve this condition, the number of linearly independent measurements is at least equal to the number of states N_s. To make the area observable with the minimum number of measurements for the estimation, we will work with topological observability, where an analysis of the type and location of measurements is required as mentioned in [12]. A condition to obtain this observability is the formation of a network tree that only considers power flows existing in interconnected lines or transformers. Using graph theory, we can represent the area as an undirected connected graph so that between nodes or vertices (area nodes) there is a path or edge (transmission lines or transformers). The minimum spanning tree consists of an undirected connected graph that contains no cycles. This tree will connect all the nodes or vertices of the graph [13]. Further, each pair of vertices will be connected by a single edge, and a tree with n vertices there are exactly $(n-1)$ edges [13].

Considering the representation of each area as a graph without edge weighting, we will proceed with its mathematical representation using the adjacency matrix, which is a square matrix of order n corresponding to the number of nodes. The matrix represents all the existing connections between the nodes in Eq. (52), where the value of d_{km} is equivalent to zero if there is no connection between the nodes or one if the connection exists. To find the minimum spanning tree, we will form a final set $Cfin$ that contains the reference node and an initial set $Cini$ that contains all the nodes except the reference node. The algorithm starts with the analysis of $Cfin$ and the connections it has with the rest of the nodes in $Cini$ using M_{ady}. The connections are chosen successively in an arbitrary way. By choosing the connection v_i the node N_i is added to $Cfin$ while it is removed from $Cini$, when $Cini$ is empty we will have obtained the minimum spanning tree, the equation that relates this principle is shown in Eq. (54), where N_i represents the new node that must be connected to the tree if the adjacency matrix between the nodes of the area relates them and v_i the vector that contains the pair of nodes of the chosen connection.

$$M_{ady} = d_{km} \tag{52}$$

$$\begin{cases} Cini = N_n \quad N_{reference} \notin N_n \\ Cfin = N_{slack} \end{cases} \tag{53}$$

$$Tree_i = \begin{cases} Cfin_i + N_i \\ Cini_i - N_i \quad , Mady_{km} = 1, N_i \notin Cfin_i \\ v_i = (k, m) \end{cases} \tag{54}$$

4 Distributed State Estimator Implementation

The implementation of the formulated estimator is done by programming a script in Matlab to obtain an automatic processing. The stages use input files (.xlsx files) of measurements, tolerance and area data. The stages of state estimator development described below are implemented in the flowchart in Fig. 3, where the minimum spanning tree stage uses Eq. (54). The iterative process of state estimation is repeated as long as the value of the differential is greater than the tolerance input.

1. **System Representation.** In order to perform the state estimation, it is necessary to establish the system model. In the electrical system, the model corresponds to the information of the electrical network that includes its nodes and interconnection lines.
2. **System Decomposition.** The system is fractioned in several areas, generated by the computational capacity, regional boundaries, etc. In the fraction process, the interconnection lines between areas (data and measurements) are disregarded. Also, for each area, there is a reference node where must exist a PMU. The slack node is the reference node for its corresponding area.
3. **Node Reorganization.** It consists of naming the nodes of the area with numerical values according to the number of nodes that actually exist in that area. This step is performed to avoid problems in the state estimation, because once the electrical system is divided in areas, the nomenclature of the buses or nodes is not affected and they use their real names.
4. **Minimum Spanning Tree Formation.** It consists of the formation of the minimum spanning tree over a certain area of the system using Eq. (54).
5. **Search for Useful Measurements.** It consists of searching the power flow in the SCADA information according to the connection vector of the Eq. (54) and the phasor voltage measurement over the reference node of the area.

6. **Admittance Matrix Formation.** It consists in the formation of Y_{bus} using Eqs. (50) and (51). The estimator requires as input the characteristic area values such as line impedances, impedances in transformers and their taps values.
7. **State Vector Determination.** It consists of the calculation of the number of states by means of Eq. (55) and the formation of the vector of states with initialization in flat profile Eqs. (56).

$$N_s = 2N_b - 1 \tag{55}$$

$$E = \left[\Theta_2 \; s \; \Theta_n \; V_{reference} \; s \; V_n\right]^T = \left[0 \; s \; 0 \; 1 \; s \; 1\right]^T \tag{56}$$

8. **States Computation.** It consists of the determination of $h(x)$ Eq. (20), $H(x)$ Eq. (21), $G(k)$ Eq. (8), R^{-1} Eq. (3). Equation (11) is used for the state computation, and the minimum value Δx is checked to repeat the estimation process.

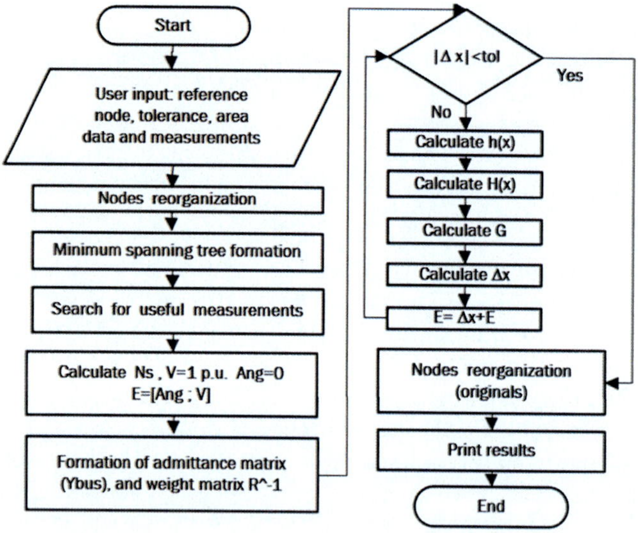

Fig. 3. Flowchart of the implemented distributed state estimator.

5 Test and Results

5.1 Operating Scenario Description

The validation of the proposed estimator will be performed on the NewEngland Power System (IEEE-39 bus) case study, which consists of 39 buses, 10 generators, 19 loads, 12 transformers and 34 lines. Figure 4 shows the case study diagram. The nominal operating values of the system 60 Hz and 345 kV. For the nodes at different voltages, it is assumed 138 kV at node 12, 230 kV at node 20 and from 30–38 16.5 kV. Only the balanced representation of the network (positive sequence) is considered in the model.

We proceed with the division of the system under the criterion of reducing the computational load in three areas. The areas are divided using the system decomposition literal of the model described above. Figure 4 shows the three areas of the system. The area data as well as the measurements, obtained from the model implemented in PowerFactory, are inserted into the distributed estimator of the area control center in Excel files. PMU and power flow measurements are in p.u. and radians and comply with the values of standard deviation of the error of the measurements voltage 0.0002, angle 0.0001 and power flow 0.02 that correspond to constant values that give good results in the literature [14].

5.2 Testing Procedures

To ensure the correct operation and validation of the proposed estimator, we proceed to the estimation in the case study, the data and measurements are obtained from the IEEE-39 bus system implemented in PowerFactory. This information is saved in the input files for the estimator. To validate the estimator, its results are compared with the real values provided by the specialized software for power system analysis (PowerFactory).

The most important concern when evaluating the quality of the estimator is the estimation accuracy. To evaluate it, we will use the root mean square error (RMSE) indicator. RMSE uses Eq. (57), where the errors are squared before being averaged giving a higher weight to large errors, this is useful in systems where large errors are undesirable as in our case.

$$RMSE = \sqrt{\frac{\sum_{i=1}^{n}(Predicted_i - Observed_i)^2}{n}} \tag{57}$$

5.3 Results

The results obtained by the proposed decoupled distributed state estimator (DDSE) are compared with the PowerFactory (PF) values and are shown in Table 1 considering voltage values in per unit and angle in degrees. The state estimation performed in the case study, shows a $RMSE$ of 0.00031 for voltage and 0.005 for angles, after being calculated by Eq. (57) using the estimated values of the three areas in addition to considering the total number of nodes in the system.

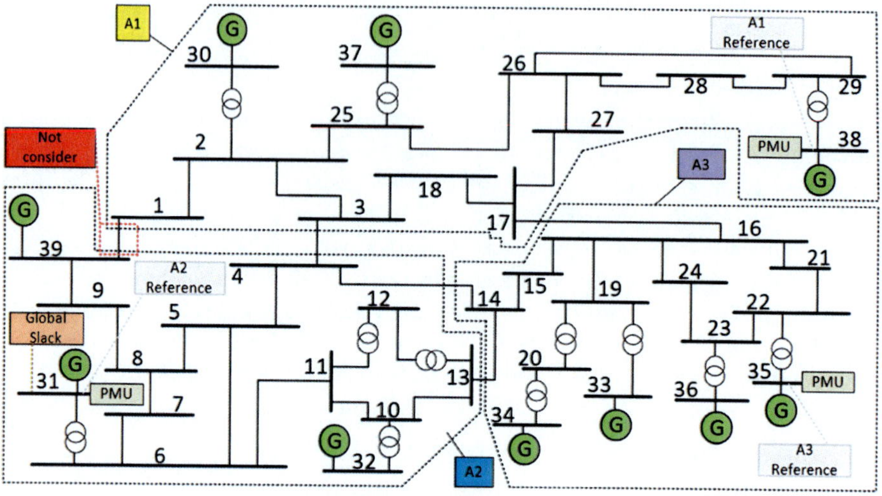

Fig. 4. IEEE 39-bus system.

Table 1. Comparison between State Estimator and PowerFactory.

Area 1					Area 2					Area 3				
	PF.		DDSE.			PF.		DDSE.			PF.		DDSE.	
N	V	A	V	A	N	V	A	V	A	N	V	A	V	A
1	1,036	−8, 66	1,035	−8, 65	4	0,955	−9, 76	0,955	−9, 76	14	0,961	−7, 6	0,961	−7, 59
2	1,019	−5, 74	1,019	−5, 73	5	0,954	−8, 63	0,954	−8, 62	15	0,969	−7, 74	0,968	−7, 73
3	0,991	−8, 72	0,991	−8, 71	6	0,955	−7, 89	0,955	−7, 88	16	0,988	−6, 07	0,988	−6, 06
17	0,992	−7, 3	0,992	−7, 29	7	0,947	−10, 31	0,947	−10, 30	19	0,99	−0, 27	0,989	−0, 26
18	0,991	−8, 31	0,990	−8, 30	8	0,948	−10, 86	0,948	−10, 85	20	0,987	−1, 26	0,987	−1, 25
25	1,028	−4, 25	1,028	−4, 25	9	1.008	−10, 61	1,008	−10, 60	21	0,995	−3, 5	0,995	−3, 49
26	1,018	−5, 45	1,017	−5, 44	10	0,962	−5, 08	0,962	−5, 07	22	1,021	1,21	1,021	1,21
27	1	−7, 54	0,999	−7, 53	11	0,959	−6, 03	0,958	−6, 02	23	1,02	0,98	1,020	0,98
28	1,019	−1, 73	1,019	−1, 72	12	0,939	−5, 99	0,939	−5, 98	24	0,996	−5, 95	0,996	−5, 94
29	1,021	1,19	1,020	1,19	13	0,96	−5, 84	0,960	−5, 83	33	0,997	4,93	0,997	4,93
30	1,048	−3, 31	1,047	−3, 30	31	0,982	0	0,981	0,00	34	1,012	3,92	1,012	3,92
37	1,028	2,55	1,027	2,55	32	0,983	2,82	0,983	2,82	35	1,049	6,18	1,049	6,18
38	1,026	8,27	1,026	8,27	39	1,03	−10, 35	1,029	−10, 34	36	1,064	9	1,063	9,00

6 Conclusions

A decoupled distributed state estimation formulation that can be applied to an electrical power system has been presented.

The low values of $RMSE$ show that the estimated values are approximate to the actual values allowing to verify the correct performance of the implemented distributed state estimator and validates its formulation. Further, the system division can be arbitrary since the query information or data exchange between areas are not required if there is at least one PMU in every area.

The computational effort for the area processor is reduced because the estimation is accomplished with a reduced number of measurements but also observability of each area is guaranteed.

Results demonstrate the feasibility of this proposal that could be a quick way for providing the current state-variables to distributed control algorithms implemented in the power systems.

References

1. Karger, C., Hennings, W.: Sustainability evaluation of decentralized electricity generation. Renew. Sustain. Energy Rev. **13**(3), 583–593 (2009)
2. Dasgupta, K., Swarup, K.: Distributed state estimation in power systems. In: Electrical Power and Energy Systems MCDES 2008, vol. 33, pp. 569–576. Elsevier (2011). https://doi.org/10.1016/j.ijepes.2010.12.010
3. Chen, Y., Kong, X., Yong, C., Ma, X., Yu, L.: Distributed state estimation for distribution network with phasor measurement units information. In: 10th International Conference on Applied Energy, pp. 4129–4134. Yinliang Xu, Hong Kong China (2019)
4. Gamboa, S.: Methodology for the design of a large-scale integrated WAMPAC system based on a distributed control architecture. Ph.D. thesis, National University of San Juan, Argentina (2018)
5. Pau, M., Ponci, F., Monti, A., Sulis, S., Muscas, C., Pegoraro, A.: An efficient and accurate solution for distribution system state estimation with multiarea architecture. IEEE Trans. Instrum. Measur. **66**(5), 910–919 (2017). https://doi.org/10.1109/TIM.2016.2642598
6. Soria, D., Gamboa, S.: Distributed static state estimator for monitoring and control of electrical power systems. Energia **13**(1), 43–53 (2017). https://doi.org/10.37116/revistaenergia.v13.n1.2017.6 https://doi.org/10.37116/revistaenergia.v13.n1.2017.6 https://doi.org/10.37116/revistaenergia.v13.n1.2017.6
7. Eghbali, O., Kazemzadeh, R., Amiri, K.: multi-area state estimation based on PMU measurements in distribution networks. J. Oper. Autom. Power Eng. **8**(1), 65–74 (2020)
8. Ren, Z., Chen, Y., Huang, S., Heleno, M., Xia, Y.: A fully distributed coordination method for fast decoupled multi-region state estimation. IEEE Access **7**, 132859–132870 (2019). https://doi.org/10.1109/ACCESS.2019.2941386
9. Macii, D., Fontanelli, D., Barchi, G.: A distribution system state estimator based on an extended Kalman filter enhanced with a prior evaluation of power injections at unmonitored buses. Energies **13**(22), 6054 (2020)
10. Hernandez, M.: State Estimator Based on an Weighted Least Squares in Power Systems Using The Newton Method, 1st edn. IPN, Mexico (2009)
11. Jin, T., Shen, X.: A mixed WLS power system state estimation method integrating a wide-area measurement system and SCADA technology. Energies **11**(2), 408 (2018)
12. Angeles, M.: State Estimation in Distribution Networks. MS. Thesis, Sevilla University, Spain (2016)
13. Acosta, B., Montoya, L.: Elements of graph theory and the evasiveness conjecture. Degree thesis, Tolima University, Ibague-Colombia (2018)
14. Rincon, R.: State estimation of a power system using synchronized phasor measurement and evaluation of its implementation in the Colombian transmission system. MS. thesis, National University of Colombia, Bogota (2013)

Object Detection and Tracking Based on Artificial Vision for a Single Board Computer (SBC)

Bryan G. Mosquera[1], Bryan G. Castelo[1], Henry P. Lema[2], Iván D. Changoluisa[2], Patricio J. Cruz[1](✉), and Esteban Valencia[2]

[1] Facultad de Ingeniería Eléctrica y Electrónica, Quito, Ecuador
{bryan.mosquera,bryan.castelo,patricio.cruz}@epn.edu.ec
[2] Facultad de Ingeniería Mecánica, Escuela Politécnica Nacional, Av. Ladrón de Guevara, Quito, Ecuador
{henry.lema01,david.changoluisa,esteban.valencia}@epn.edu.ec

Abstract. In the field of computer vision, detecting and tracking objects is an area on demand.-For this reason, algorithms specialized in tracking any object have been developed. However, those algorithms are unable to initiate the detecting process automatically since users are required to manually draw a bounding box. A potential solution is to merge an object detection neural network with a tracking vision-based algorithm. Therefore, this research proposes an algorithm developed to enhance the capabilities of KCF, an object tracking algorithm, by combining it with SSD MobileNet V2, a neural network for object identification. The proposed algorithm is developed so it can be executed on a Single Board Computer (SBC). Hence, it is possible to run and process real-time video on a Raspberry Pi 4.

Keywords: Object detection · Object tracking · SSD MobileNet · KCF · Raspberry Pi

1 Introduction

In computer vision, as in other related research areas, several alternatives can be applied to solve a problem. For example, if an object needs to be extracted from a image, it is discriminated based on its shape, color, area, or by highlighting its edges. If an application is required to identify a particular object, it is possible to use matching and detection algorithms, Haar cascades, and Neural Networks (NN). In case it is imperative to follow a mobile object in a video, specialized tracking algorithms are used such as Boosting, TLD, KCF among others [12].

The authors gratefully acknowledge the financial support provided by the Escuela Politécnica Nacional for the development of the project PIE-DIM-VLIR-2020 "Real time volcano monitoring for early eruption prediction using unmanned aerial vehicles and image processing methods".

M. Botto-Tobar et al. (Eds.): *Latest Advances in Electrical Engineering, and Electronics*, LNEE 933, pp. 154–167, 2022.
https://doi.org/10.1007/978-3-031-08942-8_12

In fact, one of the main flaws in tracking algorithms is their lack of resilience. In other words, they are unable to restart the tracking process by themselves when the tracked object is lost or occluded. Furthermore, some algorithms send an alert signal of missing object or, in some cases, they may provide wrong tracking information [5]. Thus, the user generally has to redraw the bounding box around the object to track it again.

Although there are several computer vision approaches that complement object tracking to overcome the issues described previously, this research focuses on object identification based on NN and tracking algorithms since they are able to complement each other. A NN is capable of identifying a particular object and then draws a bounding box surrounding it on the image. This feature could make the process of tracking any object much easier and automatic. In the case of an object occlusion, the alert signal of a tracking algorithm can trigger the NN in order to detect back the object in the video and retake the tracking.

In portable applications based on mobile robots several constrains must be taken into account such as available space, payload capacity and power consumption. Thus, single board computers (SBCs) are generally preferred for implementing computer-vision based applications over these platforms. However, the use of Convolutional Neural Network (CNN) is likely to overload the computational capacity of the SBC.

Tiny Yolo and SSD MobileNet algorithms are examples of optimized CNN for SBCs [2]. These algorithms reduce their accuracy compared to their full versions in order to increase execution speed. However, a SBC fully focused on running CNN-based algorithms struggles to execute them due to the limited computational capacity. On the other hand, using an object tracking algorithm as a complement reduces the computational requirements and power consumption. Consequently, the integration of these two complementary computer-vision algorithms is an alternative to lighten up the computational workload that an SBC has to handle for monitoring and tracking a predefined object.

The article is organized as follows: Sect. 2 presents a short review of potential candidates for implementing the proposed algorithm and justifies the selection. Section 3 details the main components of the proposed approach with emphasis on the software components and the algorithm design. Tests and results are shown in Sect. 4. Finally, concluding remark are presented in Sect. 5.

2 Software Review

Computer Vision focuses on processing and analyzing images automatically in order to resemble human vision [6]. For achieving this, there are many different algorithms working alongside several techniques capable of subtracting real-world information from images or video stream. Afterwards, the data is processed to execute a required action [8]. This work is based on two types of computer-vision algorithms: neural networks for object identification and object tracking algorithms.

2.1 Neural Networks for Object Identification

The Neural Network (NN) is a discriminative algorithm composed of many layers between input and the output nodes. The image data is fed to the NN, then this structure identifies an object defined in its code during the implementation process, then, it draws a bounding box around the object detected [4]. Even though these networks demand a considerable amount of time during the training process, they provide an accurate and fast response during execution stage. Therefore, they are widely used in different applications. A brief comparison of algorithms based on NN is shown in Table 1, where their lag and accuracy are considered.

Table 1. Neural networks results comparisons based on experiments in [1]

Model	Lag (ms)	Accuracy percent
Tiny Yolo V2	515	87.57%
Faster RCNN ResNet50	20018	99.33%
SSD MobileNet V2	438	91.90%

Tiny Yolo. This model is based on Yolo (acronym for You Only Look Once), a detecting object network that handles classification process as regression, making this model fast. The network divides the image in grids, then predicts the object and gives its accuracy as a class probability. Comparing Yolo to other NNs, its accuracy is slightly lower [3]. Tiny Yolo is the lighter version of Yolo which also works as an object detection algorithm capable of working real-time and showing better performance in computers with a dedicated graphic card. This network increases its speed by decreasing its accuracy [7]. Moreover, Tiny Yolo V2 is the updated version of this type of algorithm, which provides higher speed and accuracy, replacing its predecessor in most applications.

Faster RCNN. This object detection network is known for its accuracy and even being the faster version of R-CNN (Region Based Convolutional Neural Networks). However, it is not as fast as the others, (see Table 1). This network is divided in two main stages: the first one generates the boxes surrounding the objects and the other classifies and predicts the objects [9].

SSD. The Single Shot Detector (also known as SSD) is based on a single deep neural network with a structure of organized layers which draw bounding boxes around objects in order to obtain feature maps [9], making its execution time faster than other methods. A speed and accuracy comparison between Tiny Yolo, Faster R-CNN, and SSD, shows that SSD models based on MobileNet exhibit enhanced speed and accuracy. SSD MobileNet design improves the efficiency of the model and allows an on the moment usage. Furthermore, this network is widely used in several portable applications due to its lightweight and fast operation. After analyzing the aforementioned features, this network is adopted for the current study.

2.2 Tracking Algorithms

These algorithms are based on tracking the pixels inside a set of frames in a video. They estimate the movement of the selected object in order to forecast its future position. However, it stops working if the tracking object goes out of the scene range for some time or is occluded by another object. Therefore, these algorithms can return false positives results [4]. This is the main issue to be addressed in this work.

In order to choose the most suitable algorithm for this work, three different options (TLD, Boosting, and KCF) were evaluated. The metrics fps, occlusion, and out-of-view were considered for the benchmark shown in Table 2.

Table 2. Tracking algorithms results comparisons based on experiments in [10]

Parameters	TLD	Boosting	KCF
fps	18.77	30.20	230.26
Occlusion	26.7%	31.8%	36.1%
Out-of-view	26.2%	39.3%	47.0%

Tracking-Learning-Detection (TLD). TLD is a framework designed to follow an unknown object during an indefinite period of time. This algorithm consists of three stages that work in conjunction with each other: Tracking (estimates the position of the object between consecutive frames), Detecting (scans and compares current data with the previous frame in order to find the object, in some cases false positive results are generated), and Learning (analyzes the behavior of previous stages to predict and avoid future errors) [4,15].

Boosting Tracker. The Boosting algorithm is one of the oldest of its kind, its functionality is based on Haar cascades (recognition algorithms that work over each pixel of a frame and requires rigorous training) [10]. Hence, its performance is inefficient and slow. For this reason, this algorithm is rarely used. However, it is still employed as a start point for benchmarking.

Kernelized Correlation Filter (KCF). This algorithm is suitable for applications that analyze real-time videos. Its low power and resource consumption make this algorithm the most suitable for embedded devices (SBC) [13]. The

KCF tracking algorithm is based on discriminating learning methods, implementing Fast Fourier Transforms and Inverse Fourier Transforms (FFT and IFFT), which increase its efficiency and speed of calculations. However, it does not work properly when the object is completely lost of the scene [10].

The algorithm starts its process after enclosing the chosen object inside a bounding box on the first given frame, then KCF discriminates the object from the background by evaluating its position on the subsequent frames [14]. Finally, when the object is lost, due to an occlusion or for being out of scene, the algorithm triggers a *"Missed-Object"* flag.

3 Methodology

To illustrate the proposed method, Fig. 1 depicts the main components of the system and the connection between tracking algorithm and the NN. Moreover, it is necessary to check the hardware requirements and the algorithm design in order to merge object detection and tracking methods.

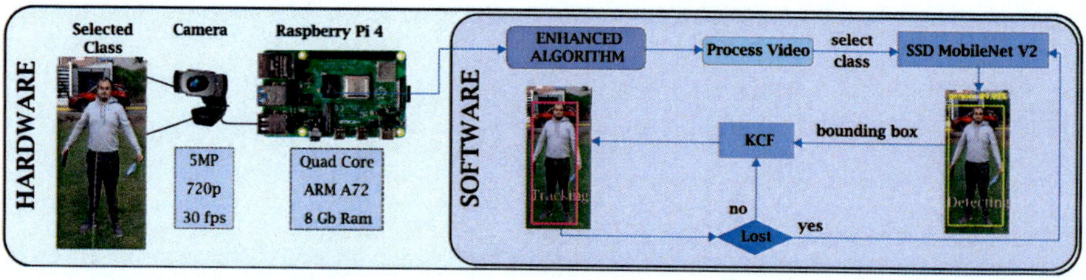

Fig. 1. Main components and algorithm connection of the proposed approach.

3.1 Hardware Analysis

The hardware required for this research has to meet the minimum requirements to run the proposed algorithm in a Single Board Computer (SBC), which is a complete computer with all its components in one circuit board, this features make it portable, small and energy efficient. Consequently, a Raspberry Pi and Jetson Nano computers are considered based on their characteristics listed in Table 3.

After analyzing the specifications of these micro-computers, the Raspberry Pi 4 with 8 GB of RAM is chosen as the SBC for this research. This equipment is characterized by its wireless connectivity, availability of accessories and RAM capacity. Additionally, it is compatible with several operating systems, supported by an active community and also its 64 bits Quad core 1.5 GHz Cortex-A72 processor can handle different types of tasks [11].

Table 3. Differences between Raspberry Pi 4 and Jetson Nano [11]

Characteristic	Raspberry Pi 4	Jetson Nano
CPU	Cortex-A72@1.5 GHz	Cortex-A57@1.43 GHz
GPU	Broadcom VI	128-core NVIDIA Maxwell
Memory	1,2,4,8 GB LPDDR4	2,4 GB LPDDR4
Wireless	Dual-Band 802.11ac	None
Price	54 $	100 $

3.2 Algorithm Design

In order to develop this proposed algorithm the artificial vision specialized library OpenCV was used, working hand to hand with Python V3.7 programming language since thanks to its versatility, it allowed the employment of OpenCV and other libraries were necessary to complete this task. The main components and more explanation about the algorithm are found in the sections bellow.

Neural Network Selection. In object detection networks, the number of alternatives are reduced because the Raspberry Pi 4 has to handle real-time video processing. Thus, the decision has to be made between Tiny-Yolo V2 and SSD MobileNet V2 because Faster RCNN could not even be executed in an SBC. In fact, both algorithms are the most used for detecting objects in a Raspberry Pi 4 [2].

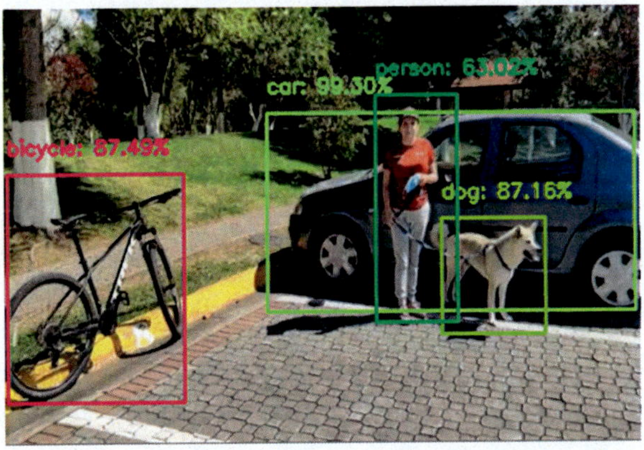

Fig. 2. Example of SSD MobileNet V2 in a Raspberry Pi 4 showing four different classes: bicycle, car, person and dog.

Therefore, tests were carried out using a Raspberry Pi 4 in order to find the performance of these NNs. It was necessary to make a data set containing 120 pictures with 262 objects total. These pictures were taken indoors and outdoors with some of the most common classes that can be identified, *e.g.*, bicycle, bottle, bus, car, dog, motorbike, person, etc. These pictures were joined to create a video being processed inside the Raspberry Pi 4. Table 4 summarizes the main results of those tests. The results revealed that SSD MobileNet V2 is three times faster, backed in a lower amount of fps of this network and more accurate compared to Tiny-Yolo V2 as it is shown in the accuracy percent. Some of them working with SSD MobileNet V2 neural network on a Raspberry Pi 4 are shown in Fig. 2. However, the limitation of the SSD MobileNet V2 for object tracking applications is that in each frame it presents the information of all the objects (class and trust) without having a system that predicts where a specific object is placed in the next frame. Further results are explained in Sect. 4. For example, the processing load for the CPU is higher when using SSD MobileNet, (see Table 7), making it run slow and increasing the chip's temperature. Even though, it exhibits an increased performance compared to Tiny Yolo V2.

Table 4. Raspberry Pi 4 neural networks test

Parameters	Tiny Yolo V2	SSD MobileNet V2
fps	1	3
Accuracy	56.95%	70.45%

Tracker Selection. In order to choose the right algorithm for the application, it was analyzed the performance of the two more promising algorithms: TLD and KCF. Due to its low performance, see Table 2, Boosting Tracker is not considered.

Table 5 provides a comparison of the two algorithms executed in a Raspberry Pi 4. The test was carried out by processing outdoor ten seconds long videos where a person appears and goes behind an object and then re-appears. The results in Table 5 were obtained by analyzing the time the person was tracked versus the total time it was is the video scene. The difference in fps between KCF and TLD is because the TLD has to complete all the three phases including the detection and learning which may not be accurate but consumes more resources compared to KCF.

Table 5. Raspberry Pi 4 tracking algorithms test

Parameters	TLD	KCF
fps	10	29
Tracking object	67.25%	70.30%
False positives	34.73%	5.12%

TLD allows tracking, detecting, and includes learning capabilities. It is capable of recovering from occlusion even so, it returns a considerable number of false positives. On the other hand, KCF is faster than TLD, but it is not able to recover from occlusion. Indeed, it triggers a "Missed-Object" flag when it stops tracking an object. Therefore, KCF was selected because it is possible to use this flag signal as a trigger for the neural network in order to restart the tracking process. Additionally, it showed less false positive percentage than TLD, see Table 5.

3.3 Algorithm Walk-Through

In order to track an object using KCF, a bounding box has to be drawn around for setting the initial parameters, then the enter key is pressed to start the process. These steps have to be done manually by a user capable of identifying the object to be tracked, as it is shown in Fig. 3. Something to be considered is that if the resolution changes, the pixel density changes as well. Therefore, if the area inside the bounding box is larger in pixels the processing speed decreases, because the tracking algorithm has to work with more information.

Fig. 3. Drawing the bounding box around the object to track.

Consequently, to initiate this process automatically, it is required to choose the class of the specific object that the algorithm is going to track before executing the code. Otherwise, it will start tracking the first detected object that appears in the video. The several classes that this proposed algorithm is capable of tracking are based on those that SSD MobileNet V2 is able to identify. Thus, it can track the objects listed as follows: airplane, bicycle, bird, boat, bottle, bus, car, cat, chair, cow, dining table, dog, horse, motorbike, person, potted plant, sheep, sofa, train, and TV monitor. This feature of choosing between several available classes, lets the algorithm be versatile and adaptable for different applications that the final user may require.

Fig. 4. Getting only the selected class to track.

The process starts by selecting the class to be tracked. Next step consists in obtaining the video stream and resizing it. In all tests, videos were resized to 480×320 pixels since the network was trained for a similar resolution and lowering this value could change its accuracy. Then, the video stream is processed by the NN in order to identify an object when it fulfills the trust value set. If this condition is not met, the NN keeps analyzing the video. In the proposed algorithm, it is possible to subtract and only validate the data of the selected class, totally ignoring the others. This step is illustrated in Fig. 4. The identified object is delimited by a bounding box which result values are designated x, y, width and height, which make it possible to draw this box, see Fig. 5. In this case, the values are $x = 102$, $y = 182$, width $= 233$ and height $= 319$. These values are used as input parameters for setting the tracking algorithm KCF. Then, KCF updates the values of the bounding box. Algorithm 1 details the proposed approach for tracking an specified object.

In the proposed solution, the tracking stage keeps working, even if suddenly an occlusion occurs or the object is lost, the proposed algorithm's resilience let it find it again. Furthermore, this event triggers the NN to restart the object detection phase and search for an object of the selected class. An example of an experiment showing this feature is given in Fig. 6 in cases of selecting a person as the class to be tracked. At first, SSD MobileNet recognizes the person, sends the bounding box parameters to set the tracking algorithm, then KCF starts the tracking process until the person goes out of the scene by an occlusion. Afterwards, SSD MobileNet searches for the selected class again, when the person re-appears, the NN is able to detect her again and set the KCF parameters to restart the tracking process.

Fig. 5. Example of bounding box parameters (x = 102, y = 182, width = 233 and height = 319).

Algorithm 1: Proposed algorithm for tracking an specific object

---

```
Result: A selected object will be tracked
Initialize: Chose the object class to be track;
while cam_frame ≠ None do
    resize video;
    NNinput(cam_frame);
    /* This part identifies the selected object                */
    while detected_object ≠ 0 do
        Subtract selected object class;
        if selected_object ≠ 0 then
            Draw bounding box;
            Sent bounding box parameters to start KCF;
            Break;
        end
    end
    /* SSD draws bounding box, continue tracking                */
    while Object being tracked do
        Continue tracking;
        if Object lost then
            Break;
        end
    end
end
```

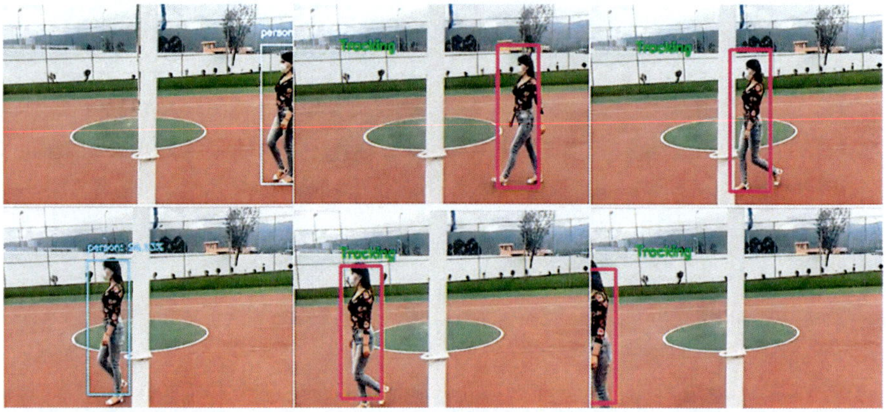

Fig. 6. Images showing the algorithm recovery when an object occlusion occurs (Test case: 1 person).

4 Test and Results

As it was explained in Sect. 2, the first step consisted on selecting a NN able to be executed smoothly in a Raspberry Pi 4. The best results from the tests made were obtained with the SSD MobileNet V2. However, when it runs the process for a extended period of time, the Raspberry Pi 4 starts to heat up, see Table 6. Consequently, the SBC temperature turned out to be an issue since high temperatures make the chip under-perform and decreases its processing speed.

In addition, while executing only the neural network, the CPU load is more than twice higher than using only the tracking algorithm as it is shown in Table 7. Thus, using tracking algorithms is faster and needs less processing power but they always require setting their initial parameters or as known as draw the bounding box.

Table 6. Raspberry Pi 4 temperature test

Temperature CPU test raspberry Pi 4			
Time	**SSD MobileNet**	**KCF**	**MobileNet + KCF**
0 min	36 °C	36 °C	38 °C
2 min	51 °C	42 °C	43 °C
5 min	57 °C	46 °C	46 °C
10 min	62 °C	51 °C	51 °C

Based on the results, KCF is the phase of the proposed algorithm that runs most of the time. Therefore, the CPU temperature and CPU load are lower than using only the neural network all the time. Also, by combining SSD MobileNet

Table 7. Raspberry Pi 4 CPU load and RAM test

Time	SSD MobileNet		KCF		MobileNet + KCF	
Seconds	CPU	RAM	CPU	RAM	CPU	RAM
0 sec	3%	2.16%	3%	2.16%	3%	2.16%
10 sec	97%	3.11%	42%	3.19%	91%	3.11%
30 sec	90%	3.11%	42%	3.19%	42%	3.19%

with KCF, the amount of fps obtained is superior, therefore it is possible to get a fluid image even if the RAM usage is slightly higher.

In order to test the accuracy and the capabilities of the proposed algorithm, twenty-five different videos were used. Some objects of different classes are presented in the scene. Some graphic examples are shown in Fig. 6 and Fig. 7 (also accessible in YouTube). In these tests, the algorithm showed the capability of identifying only the selected class and then tracking it even when an occlusion occurs or the object goes out and re-enters in the scene. Finally, the values presented in Table 8 are the ratio of the time the object is tracked and the time it is displayed on the scene.

Table 8. Average proposed algorithm's accuracy percent

Video 1 to 5	Video 6 to 10	Video 11 to 15	Video 16 to 20	Video 21 to 25
92.94%	89.88%	90.78%	92.60%	90.22%
Total average = 91.28%				

*The accuracy average were calculated with five videos each.

Fig. 7. Example of the proposed algorithm working with occlusion and two people

5 Conclusions

Tracking an object in a Raspberry Pi 4 using the proposed algorithm has satisfactory results. It makes the tracking process automatic because the network identifies the object, then it uses those parameters to start the tracker. Also, If the object is lost, the algorithm can find it again and track it by itself. This switching stage between the tracking and detecting algorithms makes them complement each other and by interchanging them it is possible to reduce resources consumption and improve speed due to the computational load reduction for the SBC.

Using only the neural network as a tracking method is not the right approach because its architecture is designed to identify objects rather than track them. Therefore, it consumes more computational resources than working with the specialized algorithm for tracking any object. Consequently, the temperature rises and the processing speed of the SBC slows down.

The KCF algorithm was chosen because it exhibits the best performance for this application. It is fast, accurate and as soon as it misses the tracked object it triggers a Missing-object flag in order to report a change of state. This message can be used to execute an additional process, e.g. a neural network. Although there are other specialized tracking algorithms for similar applications, this research strongly suggests the use of the KCF combined with a neural network due to the satisfactory results.

By using this approach it is possible to get a portable, small and affordable alternative for identifying objects and being able to track them in the scene, lowering the computational power needed and making it automated. Since the present study only covers the characteristics and functionalities of the proposed algorithm, in future works, the comparison with other algorithms that perform this type of combined tasks will be considered.

References

1. Alsing, O.: Mobile Object Detection using TensorFlow Lite and Transfer Learning. Master's thesis, KTH Royal Institute of Technology (2018)
2. Gunnarsson, A., Davidsson, M.: Real time object detection on a Raspberry Pi. Master's thesis, Linnaeus University (2019)
3. Huang, R., Gu, J., Sun, X., Hou, Y., Uddin, S.: A rapid recognition method for electronic components based on the improved YOLO-V3 network. Electronics **8**(8), 825 (2019). https://doi.org/10.3390/electronics8080825
4. Kalal, Z., Mikolajczyk, K., Matas, J.: Tracking-learning-detection. IEEE Trans. Pattern Anal. Mach. Intell. **34**(7), 1409–1422 (2012). https://doi.org/10.1109/TPAMI.2011.239
5. Lehtola, V., Huttunen, H., Christophe, F., Mikkonen, T.: Evaluation of visual tracking algorithms for embedded devices. In: Sharma, P., Bianchi, F.M. (eds.) SCIA 2017. LNCS, vol. 10269, pp. 88–97. Springer, Cham (2017). https://doi.org/10.1007/978-3-319-59126-1_8

6. Meré, J.M.: Técnicas de visión por computador para la reconstrucción en tiempo real de la forma 3D de productos laminados. Ph.D. thesis, Biblioteca de la Universidad de Oviedo, Oviedo (2009)
7. Oltean, G., Florea, C., Orghidan, R., Oltean, V.: Towards real time vehicle counting using YOLO-tiny and fast motion estimation. In: 2019 IEEE 25th International Symposium for Design and Technology in Electronic Packaging (SIITME), pp. 240–243, October 2019. https://doi.org/10.1109/SIITME47687.2019.8990708
8. Patel, P., Thakkar, A.: The upsurge of deep learning for computer vision applications. Int. J. Electr. Comput. Eng. (IJECE) **10**(1), 538 (2020). https://doi.org/10.11591/ijece.v10i1.pp538-548
9. Qurishee, M.A.: Low-cost deep learning UAV and Raspberry Pi solution to real time pavement condition assessment. Ph.D. thesis, University of Tennessee at Chattanooga, May 2019
10. Raghava, N., Gupta, K., Kedia, I., Goyal, A.: An experimental comparison of different object tracking algorithms. In: 2020 International Conference on Communication and Signal Processing (ICCSP), pp. 0726–0730. IEEE, Chennai, India, July 2020. https://doi.org/10.1109/ICCSP48568.2020.9182101
11. Suzen, A.A., Duman, B., Sen, B.: Benchmark analysis of Jetson TX2, Jetson Nano and Raspberry PI using Deep-CNN. In: 2020 International Congress on Human-Computer Interaction, Optimization and Robotic Applications (HORA), pp. 1–5. IEEE, Ankara, Turkey, June 2020. https://doi.org/10.1109/HORA49412.2020.9152915
12. Ullah, K., Ahmed, I., Ahmad, M., Khan, I.: Comparison of person tracking algorithms using overhead view implemented in OpenCV. In: 2019 9th Annual Information Technology, Electromechanical Engineering and Microelectronics Conference (IEMECON), pp. 284–289. IEEE, Jaipur, India, Mar 2019. https://doi.org/10.1109/IEMECONX.2019.8877025
13. Wan Park, J., Kim, S., Lee, Y., Joe, I.: Improvement of the KCF tracking algorithm through object detection. Int. J. Eng. Technol. **7**(4.4), 11 (2018). https://doi.org/10.14419/ijet.v7i4.4.19594
14. Wei, B., Wang, Y., He, X.: Confidence map based KCF object tracking algorithm. In: 2019 14th IEEE Conference on Industrial Electronics and Applications (ICIEA), pp. 2187–2192. IEEE, Xi'an, China, June 2019. https://doi.org/10.1109/ICIEA.2019.8834374
15. Xu, T., Huang, C., He, Q., Guan, G., Zhang, Y.: An improved TLD target tracking algorithm. In: 2016 IEEE International Conference on Information and Automation (ICIA), pp. 2051–2055. IEEE, Ningbo, China, August 2016. https://doi.org/10.1109/ICInfA.2016.7832157

Development of an Industrial Communication Driver for Profinet Devices

Gabriel Santos[1(✉)], Silvana Gamboa[1,2], and Ana Rodas[1]

[1] Escuela Politécnica Nacional, Quito, Ecuador
gabriel.santos@epn.edu.ec
[2] GIECAR, Escuela Politécnica Nacional, Quito, Ecuador

Abstract. This document presents the development of a low-cost industrial data server for communicating Profinet devices and Windows applications by using a database as a temporary storage space. The performance of developed data server was evaluated by integrating it with an commercial industrial Windows application, were data server demonstrated an adequate behavior. The proposed application is based largely on free software, Java and MySQL, to reduce its cost and to become it an accessible option for industries with low investment capacity. Then, this development aims to offer a low-cost alternative for implementing a supervisory control system.

Keywords: Profinet · Communication · Communication driver · Java · Database · MySQL

1 Introduction

In recent years, there is a growing interest in smart manufacturing, not only from manufacturers, but also from governments and, of course, researchers, due to its promising future in reducing energy consumption, increasing economic benefits, and enabling custom production [1–3]. In this regard, a critical component toward smart manufacturing is communication between the industrial automation and control system (IACS) and PCs that perform smart manufacturing applications. For this purpose, communication drivers are used, which enable data exchange between industrial equipment and computer systems. However, the high investment required for such software is becoming a limitation for small and medium-sized enterprises (SMEs) due to its low investment capacity [4,5].

Currently, development of industrial applications in free software has become a low-cost alternative. Thus, applications such OPC servers for information exchange between industrial devices and supervisory control application as Argos and FreeSCADA [5] have been implemented. This kind of application has allowed

The authors acknowledge support from Escuela Politécnica Nacional research grant PII-DACI-2021-01.

M. Botto-Tobar et al. (Eds.): *Latest Advances in Electrical Engineering, and Electronics*, LNEE 933, pp. 168–182, 2022.
https://doi.org/10.1007/978-3-031-08942-8_13

industries with low purchasing power to take their first steps towards digitization of their processes. National industry cannot be an exception, then a local development will facilitate access to these technologies for the small Ecuadorian industries.

On the road toward development of the proposed communication drivers, it is important to establish more common industrial protocols. In this regard, Profinet has been identified as a protocol that has a strong presence in industrial applications, because it is available through the interface integrated in programmable logic controllers (PLCs) from brands such as Siemens. In such controllers, data exchange through Profinet can carry by using open user communication instructions that can be included in the controller's programming. Such instructions allow configuration and creation of the communication link, sending of data through the link and disconnection from it.

With this background, this work is focused on developing a driver for communicating Profinet devices to Windows applications by using free and open source software. In order to enable the proposed data exchange a database is proposed as intermediary application. Then, the developed driver establishes communication between Profinet PLCs and this database in which process input data will be stored to the Windows applications can access them later from this database. Also, the use of a database as register mechanism will help later to implement a kind of process historian.

This work is organized as follow. Section 2 presents a review of Profinet protocol and details the structure of its data frame. Then, requirements for software development are exposed in Sect. 3. In Sect. 4, procedure for mapping Profinet frames over TCP is detailed. Later, communication with database is explained in Sect. 5. Section 6 describes the development and operation of server configuration interface. In Sect. 7 tests and their results are presented, and finally, the conclusions are drawn in Sect. 8.

2 Theoretical Framework

2.1 Profinet IO

Profinet was born as an evolution of Profibus DP protocol proposed by the Profinet International Foundation. This one is standardized in the normative IEC61158 and IEC61784. But it is important to emphasize that Profinet has been proposed with characteristics that allows it to have more capabilities than Profibus DP. This is because of its capability to be implemented over Ethernet 100 Mbps, which allows it a broad interoperability [3].

The expectations for Profinet in the industry 4.0 include proxy communication and APL (Advanced Physical Layer) interrupters. Profinet being used as a way of communication between controllers and the field components. For the whole understanding of the Profinet Framework we have to take in count, the Ethernet protocol, then the IPV 4 protocol as well as the TCP protocol. Also it has to be taken in consideration the MAC address as well as the IP address and ports used in each of the elements as well as the flags turned on based on the message characteristics.

2.2 Ethernet Protocol

For this case, the Ethernet header begins with the Mac address of the addressee of the message, this means 6 pairs of bits that present the unique characteristics of the device to send the message, in which the LG and IG bits are occupied. That serves to determine individual characteristics of the MAC address. After the MAC address of the recipient of the message was informed, the Mac address of the origin of the message is entered, which in this case, in the same way as in the previous addressing, occupies 6 pairs of bits. After that, the type of IP protocol is determined, which in this case is IPv4 and the padding id added [6].

2.3 IPv4 Protocol

The internet protocol in this case is IPv4, it uses multiple identifiers. First, the type of version is mentioned, there are currently two versions, IPv4 and IPv6. The IPv4 version, which is the one that will be used for the development of the project, includes a 4-byte address. The length of the header should also be mentioned after the version of the IP protocol. There is a field of differentiated services that in general cases is in default since it is not required for this process. Then the total length and the identification number are indicated. It is mentioned: the number of flags, the fragment offset, time to live, type of protocol used; in this case this is the TCP protocol. Finally the error checking and the source and destination IP addresses are obtained [6,7].

2.4 TCP Protocol

TCP is known as the Internet connection-oriented transport protocol, in this case it is applied as an Ethernet connection. For this case we have a transport in a TCP/IP architecture, it should be noted that there is a considerable difference between a TCP system and UDP system, in the case of Profinet a TCP type header is used directly. The TCP header works with the source port and the destination port, then each part of the protocol is shown: source port, destination port, frame index, segment length, sequence number, acknowledge number, header length, flags, window size, error checking, recognition sequence analysis, time stamps, TCP Payload. The flags used in the TCP header are Aknowledgment and Push [6,7].

3 Proposal for Industrial Driver Communication Implementation

3.1 Commercial Data Servers

Many industrial software brands offer data servers that can accomplish the objectives proposed in this work, but as it's mentioned above the associate cost is a limiting for some industries. Some of this commercial driver are brief described below:

- Wonderware: DAServer. Created by Wonderware as a windows aplication. It is enabled to communicate with the Siemens s7 family of Controllers.
- Matrikon: OPC server created by the Matrikon company. It works with the S7 siemens family.
- National Instruments: This driver is created to be used in the development of tools oriented to industrial control using the Labview Software. It works with Siemens [5].

Some characteristics of this data servers were taken as reference for our driver implementation. It's important to take in consideration that although these data servers usually are associated with an HMI software, they can be used with different applications as long as such applications have an standard mechanism for information exchange.

3.2 Proposed Architecture

The system was implemented for communicating Profinet PLCs with a HMI application through a database. In order to accomplish this objective, the driver takes the information from the HMI and saves this information in a relational database. After the data is stored in the database, the driver decodes this information and send it to a PLC. The process before described also works in the other way. In which the information is obtained from the PLC by the driver and therefore written in a relational database to be used by any application that requires it. The architecture that fulfill this objective is shown in Fig. 1.

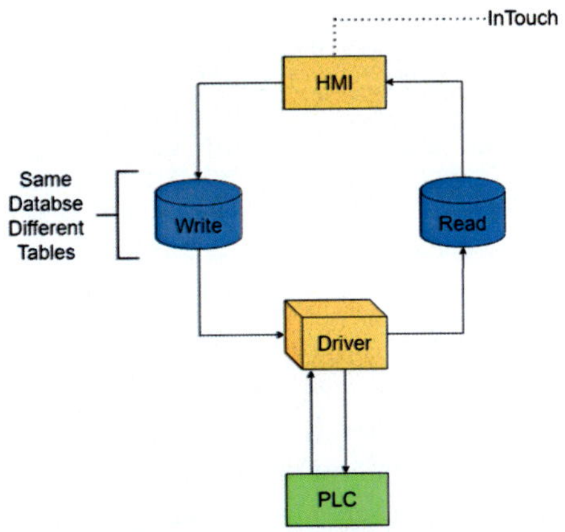

Fig. 1. Working architecture for the driver.

To develop the system it is important to take in consideration that the Siemens software in the PLCs is not developed to be used in this way. There is the reason why it is needed to program the PLC as if it was communicating with another PLC, instead of a computer.

3.3 Software Selection

Java. Java is a high-level programming language. This means that this type of programming it's quite away from the direct programming for the microprocessor of the computer. In Java programming it is required to state programming statements which are similar to a sentence in a language. These statements take net programming job and are also known as commands [9].

MySQL. The general idea of the development of this driver is to use accessible tools for it's development. This is the reason why it's used the MySQL workbench. This branch of MySQL allows the creation of a database freely. Also, it's important to take in consideration that this tool also has an interface that makes the modification and creation of database quite easily to people that doesn't know MySQL programming. Even though the creation and modification of a database in the MySQL workbench interface is relatively easy. It's important to know about the my SQL programming and how the statements in this programming work in order to develop the driver [8].

Relational Database Model. The model from a relational database is the common idea that its shown when talked about a table in a database. A relational database is made from multiple tables, named relations. Each table has their own columns and rows.

To be more clear the table can be understood as a relation. In this relation there are some attributes, which are known as the columns and there are also records whom are also known as rows. After the information is finally archived in the database. These terms are the ones used to identify data in the storage [8].

3.4 Existing Variables in the PLC and Their Management

For the use of the variables on the PLC, we have to understand how these variables work. This is because we have to know how the algorithm uses these variables in order to understand what information comes from the PLC or else goes to the PLC.

Floating Point Variables. Floating point variables are standardized by IEEE754. It has two presentations the first one is the simple precision and the second one is the double precision. Each of these representations has an specific organization of the bits included in the package. This also means that the information obtained must be analyzed bit by bit to solve the actual data received or sent. In the development of this project is only used the simple precision [5].

Integer Type Variables. We also used it integer type variables. These ones are just both for the transmission of data type int as well as data type bool. This happens because the TCP protocol used in the Profinet transmission doesn't allow to only have one bit in their whole package. this makes sense because the whole package is way too big to only send one bit at a time. Therefore, there is also a conversion algorithm that exists for the integer type of data [5].

3.5 Existing Variables in the MySQL Database and Their Management

In the proposal it is considered the implementation of the driver with a MySQL database, taking into account the requirement of a bridge between the Java programming and the HMI developer. Hence these requirements are presented its important to acknowledge the types of variables used, and their characteristics regarding the amount of bits required of each of these mentioned (Table 1).

Table 1. Variable type used in MySQL [8]

Type	Bytes
Tinyint	8 bit integer
Int	32 bit integer
Float	4 bytes floating point

The MySQL interface allows us to modify the amount of bits used in a type of variable. Therefore in order to couple the information between the data types existing in the HMI developer, there are some modifications realized in the writing process.

4 Communication Driver Implementation

The driver was developed taking as the basis the TCP communication protocol between PLC and it's adaptation to the Profinet protocol. In parallel to this analysis, it is understood how to manage this information, that is looked forward to being received or sent, in order to accomplish the architecture planned. Therefore, the communication and usage of the database is required. The driver is the central application of the whole process and manages all the information that goes through. Also, it traduces and configures the data blocks in order to make them understandable for each part of the communication.

4.1 Profinet Data Management in Siemens

To understand the workflow and the management of the data needed in the Profinet protocol. It is required to understand how the PLC in the Siemens software is programmed. It's also requires the development of the driver, as well as the configuration, to understand how the data management and storage works in the controller.

To configure Profinet communication, we should understand first, how do internal blocks of programming work in the PLC. We have two blocks, the first one that must be configured to send information to the other controller that also has a specific block that works in receiving this information. In terms of fact, when we are using the driver, we only need one block. It's only necessary to create a symbolic PLC to represent the computer under connection in the network.

A TSEND block is needed. By the creation of this block, it also created a database with the same name as well as different data blocks created specifically to watch the performance or the status of the communication. This also happens in the receipt block. For the general configuration, we use a clock, a block activator, as well as a variable generated from the configuration of the connection between the two blocks. As shown in Fig. 2. Finally, it also requires a data block that contains information to be sent or to be received.

The internal configuration of the block just requires that we choose the right parameters for the configuration as well as a connection ID. Which is created for each of the blocks generated, as well as a connection data variable that is created automatically with the connection. One of the most important steps in the internal configuration of the block is the choosing of the port that will be used to send information, this is because this port will later be used to configure the driver.

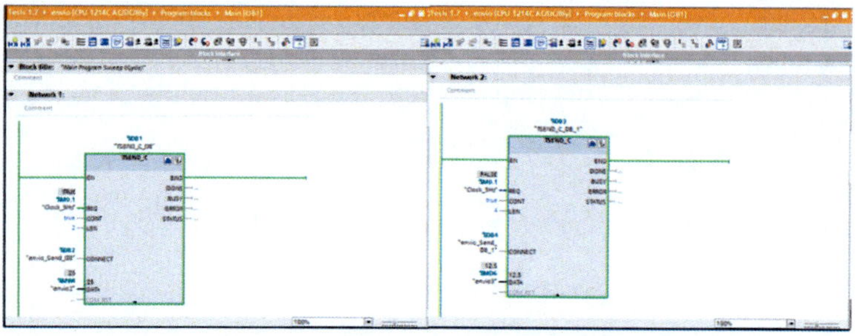

Fig. 2. Programming of a TSEND block created in the TIA portal programmin tool.

4.2 Profinet Protocol Initialization

For the understanding of the Profinet protocol, it's important to take into consideration how the communication between two devices that work with this protocol begins and continues through time. First, we have to understand the TCP protocol. There is always a search and a searcher. Therefore, we must know which controller is the searcher and which controller is the searched. This is chosen by programming the block configuration in the programming tool that Siemens offers.

Reaching the communication process. First, we have a broadcast message that comes from the device that initiates the communication, looking for the device that is also part of this communication process. It's important to take into consideration that the search that is made by the device is done by the IP address. When the other device answers to this broadcast message, it tells this device what its MAC direction is.

After the device has the MAC direction, it sends to the other device a synchronism message. This message has an active flag in the TCP protocol. Answering this message is another synchronism acknowledge message with both the acknowledgment and synchronism flags turned on. Finally, It's sent another acknowledge message but this one is not for synchronism; this one is used to finish the communication initialization. Then we have the beginning of the package sending with their own acknowledged answer. The whole process is shown in the Fig. 3.

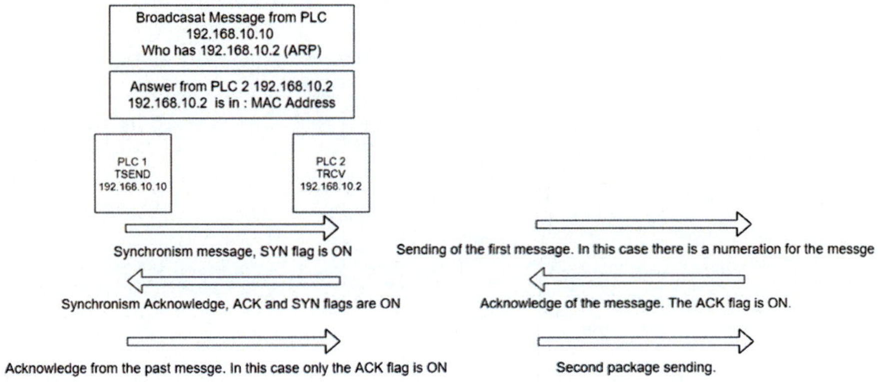

Fig. 3. Diagram showing the process of communication between devices in profinet.

4.3 Software Implementation

The implementation is done in Java language. The whole program is supposed to be able to read and write information in the PLC using the TCP and Profinet protocol. As well as, being capable of reading and writing information in the MySQL database.

Each algorithm as said before is analyzed. Therefore each of the expected results is shown, in order to be able to create an expected result of the whole system. Here are shown as well the libraries or the statements used to accomplish the tasks of each of the architecture components.

Common Statements in MySQL. For the development of this project, it's important to know first of all, how the variables in the MySQL language work, as well as the statement are used at the SQL language. therefore, here is represented a table with the most common statements from the language SQL (Table 2).

Table 2. Common statements in MySQL [8]

Statement	Usage
Select	Data search in a table
Insert	Data insert
Update	Used to modify data in a table
Delete	Deletes information in a table
Where	Condition normally in a statement that tells us where to consult data
Drop table	Table elimination
Alter table	Modify the characteristics of a column in a table

Sending Frames over TCP. To understand how the data was sent via Profinet Protocol. It was first necessary to see through a package analyzer that shows how the data frame is made. Therefore after the frame its obtained, its able to replicate this message in the Java programming. To understand in a big way how this frame is conformed, its important to take in consideration the IPv4 protocol, TCP, and Etehernet 2. The data package shown next in the Fig. 4, comes from a test made between two PLCs. This information is obtained from the wireshark tool that allows to see the trafic flow in a network.

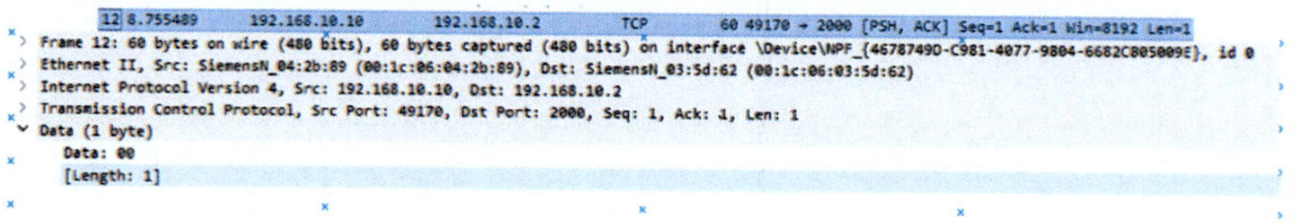

Fig. 4. Frame in profinet protocol from PLC to PLC.

In order to replicate this information in the Java programming. It's required to use the libraries: Socket, SocketAdress and InetSocketAddress. These libraries are used to create and modify a TCP frame. The most important configuration required in the programming regarding the network information is the port used in the device for receiving or sending the information, as well as the IP address from the partner in the network. The library creates a socket with the information given.

Next is shown in the Fig. 5 a frame sent from a computer to a PLC, captured with the wireshark tool. This information was taken from a test realized for the understanding of the usage of the libraries before mentioned. It's important to point out that the MAC address analyzed by the wireshark tool shows that one of the devices in the communication is a Realtek adapter (PC).

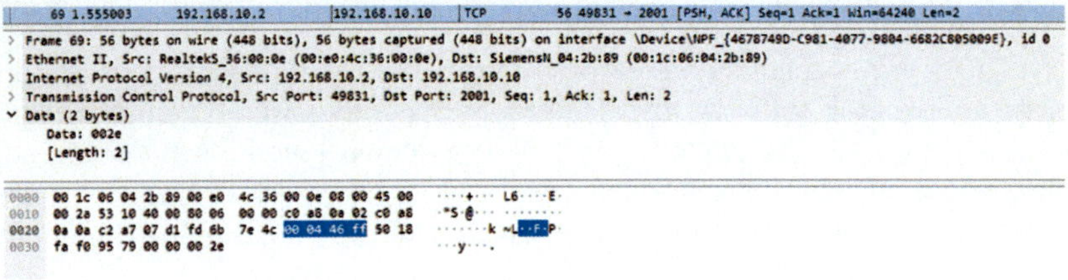

Fig. 5. Frame in profinet protocol from PC to PLC.

4.4 Driver Configuration Interface

For the configuration there were considered two configuration windows as well as two testing windows. The implementation of the testing windows works as a tool for the engineer to know where is a problem and how it can be solved. The configuration windows work directly to be the anteroom before the operator window.

Read PLC Write Database. The project considers two main windows. One of those is the "read PLC write Database" window, shown in the Fig. 6. As the name tells, the objective from this window is to take information from the PLC, then write this information in a Database. Before it was already shown how the information from the PLC is taken. After this process is successfully done, the driver creates a connection with a table from a database before created.

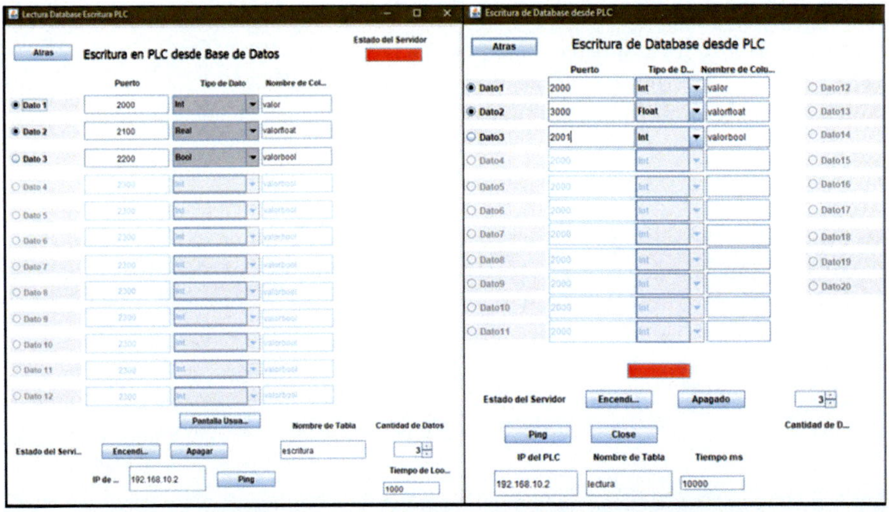

Fig. 6. Configuration window for the reading and writing driver.

When the connection is successfully done the driver runs an algorithm that takes the user requirements and adapts the table in the database in order to fulfill these. Everything that the algorithm does depends strictly from the requirements that the user configures in the window shown before. After the amount of columns was checked the driver names these columns according to the users requirements. Its important to take in consideration that before the columns have names, the algorithm gives them a number in order to just create the column. Everything considered before only happens once every time the server is turned on.

The cyclical algorithm developed happens continuously. The information obtained from the PLC is saved continuously, ruled by a timer that the user configures. The algorithm first creates a row of zeros in the database. This is done because in this way is easier to write the information in the database.

Write PLC Read Database. The second window works writing the information read from the Database to the PLC, shown in the Fig. 6. This one works as well with the algorithm before mentioned to receive information in a TCP protocol from the PLC. Before the information is sent to the PLC, the algorithm accomplishes the task of obtaining the information from the database. Therefore is required the information that helps the driver to find the data in the database.

The algorithm has been made to only look for the last row in a column in the database. This means that the only information required to obtain information is the column name, as well as the table name. The algorithm begins by creating a connection with the database. After the connection is successfully made, the algorithm uses the statement select to obtain specific information from the database.

With the name of the column, using the statement mentioned before, the algorithm obtains the information required from the user. This information is sent to the PLC via the writing TCP algorithm before shown. This whole process happens every time a timer configured by the users, says to do it.

5 Test and Results

For the analysis of the results from the driver, first is required to test the functionality of the driver and the viability of using this driver also in commercial HMI software. It is also required the testing of the system capability of being used and tested without a commercial HMI, so for this testing it shall be used the option of the testing tool also included in the software.

The functionality and the process analysis will be developed in the programming software from Siemens, known as TIA Portal.

5.1 Testing Requirements

The database in order to work shall be configured as a DNS in the computer that has the HMI. This configuration was done by a external ODBC driver that was provided by the MySQL community, this one is shown in the Fig. 7.

It's important to take in consideration the software used in both the HMI development tool, as well as the capability of both the computer that has de HMI as well the computer that has the driver running. The project was developed with this warning in mind. Therefore the driver of communication, as well as every application used were configured to work with 32 bits. Otherwise an error occurs in the HMI development.

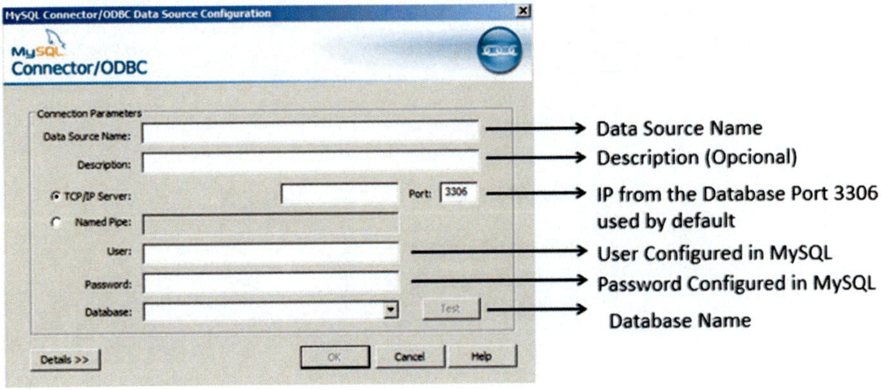

Fig. 7. Driver used to create the DNS required to use the HMI developement tool.

5.2 Lecture Process

In this process we take the information from the PLC, and then write it on the database. To manage the viability of this process, it is also developed a simple application on Intouch in a machine connected to the Network. Finally, we take the information from the database into an HMI.

For the block programming in the PLC it is required to use a clock mark, because it references the time required between each of the packages sent to the computer. One important consideration to take is that the amount of bytes sent shall be accurate depending of the type of data sent. In this case the clock mark 5 Hz. Check fig 2 for the TSEND configuration.

Application. First, a tag in the program of the PLC is modified so it can send this information via Profinet to the computer, the Port of reception is 2000 and 3000. To test that the process is working properly we can check first the testing tool of the Java application.

After checked, that the connection is working and so on to continue the configuration of the driver. Since we are working only with two packages the configuration of the system should be as presented next in the Fig. 8:

The driver is configured selecting the amount of columns wanted to be created in the Database. In this case this option is selected as 3 so the program wont have to create or add columns in order to run the algorithm. After that the Ip

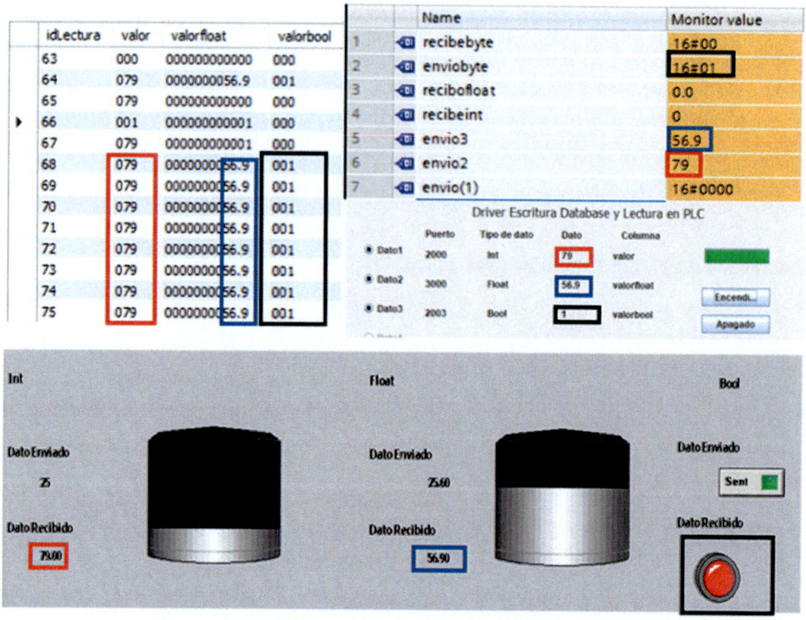

Fig. 8. Application of the driver with a simple example developed in Intouch.

configuration of the PLC is selected as 192.168.10.2 and the name of the table in the MySQL database, as well as the time required to update every part of the database.

The name of the column is also required, so it can be stored in the database. With this information the driver can modify and create the columns, so it can fit the information in the HMI developer tool.

5.3 Writing Process

In this process we take information from the Intouch application before developed and then we save this in a MySQL database. The driver takes the information in the database based on the configurations made by the user and therefore sends it to the PLC. The whole system is tested at the same time so it can be verified as a working process.

For the block programming it is required to use a clock mark, because it references the time required between each of the packages received. Consider the amount of bytes sent. The clock selected for this application is the 2.5 Hz. The ports used are the 2001 and the 2002 (Fig. 9).

Application. After the connection is checked, it's required to configure the information from the PLC. Including the IP of this one in the network. It's also required the database name as well as the columns from which the information

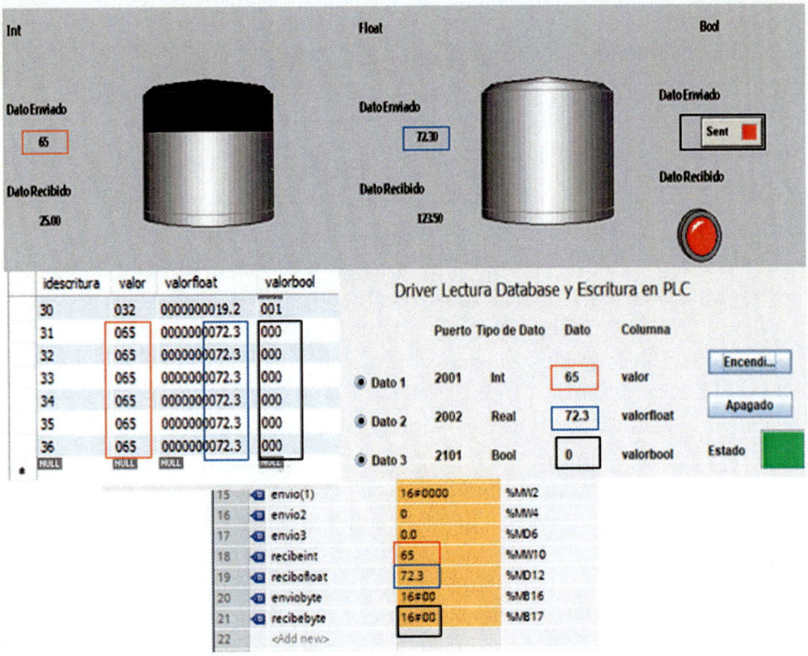

Fig. 9. Application of the driver with a simple example developed in Intouch.

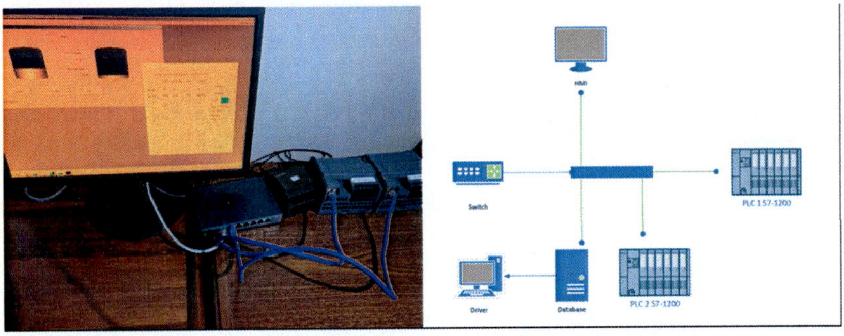

Fig. 10. Photo of the system applied.

shall be taken, this means that another requisite is the amount of packages sent from the computer to the PLC. Take as reference the Fig. 10.

The information regarding the database shall be equal to the one configured in the HMI so there would be no misunderstands of information. The column names from which the information is taken is configured as well, this part is done for each of the packages, in this case the names are *"Intdato"* and *"Floatdato"*, since these are the names of the columns selected before in the HMI and configured first in the database. A ms loop time is required so the data can be updated constantly. The time chosen is 1000 ms.

As done before after checking each individual configuration, it's able to proceed to the full implementation of the software as shown in the Fig. 11.

6 Conclusions

Design and implementation of industrial driver communication that enables connection between Profinet devices and Windows Applications has been presented. From the tests and results, it's possible to conclude that it is feasible to develop a driver for substituting some tasks of commercial industrial drivers.

The driver has shown that is qualified to obtain information from a Profinet PLC network and decode this information in order to be used by any application that has the ability to obtain data from a database.

It's concluded that the use of free and open-source software enables to develop a driver that fulfills requirements of industrial automation. This means that, with an appropriate configuration and implementation of the driver, it could be potentially used in an industrial environment as a communication tool for supervisory control applications.

It is important to highlight that the driver could be improved for a much more complex application. The development of a cloud based driver with historic applications is suggested as a nearest possible upgrade to be implemented. Of course, this proposal would requires a better security functions related to communication to the cloud.

References

1. Liu, Q., et al.: An application of horizontal and vertical integration in cyber-physical production systems. Int. Conf. Cyber-Enabled Distrib. Comput. Knowl. Discov. **2015**, 110–113 (2015). https://doi.org/10.1109/CyberC.2015.22
2. LNCS. http://www.springer.com/lncs. Accessed 4 Oct 2017
3. Wenzel, P.: Profinet - Lösungsplattform für die Prozessautomatisierung (2015)
4. Aguirre, D., Gamboa, S., Rodas, A.: Low-cost supervisory control and data acquisition systems. 4th IEEE Colombian Conference on Automatic Control as Key Support Industrial Productivity CCAC 2019 - Proceedings, pp. 1–6 (2019)
5. Maldonado, V., Gamboa, S., Trujillo, M.F., Rodas, A.: Development of an industrial data server for modbus TCP protocol. In: Botto-Tobar, M., Gómez, O.S., Miranda, R.R., Cadena, A.D. (eds.) Advances in Emerging Trends and Technologies. ICAETT 2020. Advances in Intelligent Systems and Computing, vol. 1302, 16–28. Springer, Cham (2021). https://doi.org/10.1007/978-3-030-63665-4_2
6. Wu, X., Xie, L., Lim, F.: Network delay analysis of EtherCAT and PROFINET IRT protocols. In: IECON Proceedings (Industrial Electronics Conference) (2014)
7. Comer, D.: Internetworking With TCP/IP. Internetworking Res. Exp. (2005)
8. Matthews, M., Cole, J., Gradecki, J.D.: MySQL and Java Developer's Guide (2003)
9. Schildt, H., Schildt, H.: Java: the complete reference (2007)

Information Networks

Proposal for Information Security Risk Mitigation Practices Based on a Regulatory Approach

Alejandro Andrade Mafla(✉)

Costa Rican Institute of Technology (TEC), Barrio Amón Calle 5, San José, Costa Rica
aandrade@estudiantec.cr

Abstract. Business objectives could be affected by the materialization of several types of risks, including operational, legal, or contractual, technological and information security risks. As part of the treatment and mitigation of the threats, specifically those related to regulatory and information security, the following research work offers proposed mitigation actions - adjusted to the organizational context of a company - that are required to treat the aforementioned risks. This is done by using the ISO 27002 reference guide as a tool, and two specific input variables: a) organizational risk profile and b) a regulatory environment restricted to the nature of the organization. The study is conducted for a Costa Rican financial institution dedicated to pension fund management. As a result of the analysis, mitigation practices are established, which contemplate an outline of the organizational activities and responsibilities related to regulatory compliance with the general regulation of information technology management and the law of personal data protection, providing guidelines that enrich the internal management regarding the treatment of failure modes identified during the study period.

Keywords: Control activities · Information security · Protection of personal data · Regulatory impact

1 Introduction

According to [1] information security incidents tend to grow with the development of digitalization and its benefits, where attack methods are specialized according to the industry they affect. The financial industry is not exempt from this trend. According to the study conducted by (Ponemon IBM Security, 2017), the financial industry presents the second highest cost for data security breaches, distributed in criminal attacks, system errors, and human errors. In order to face this reality, Costa Rica has three representative regulations that seek to protect the information of its inhabitants. The first one corresponds to the general regulation of information technology management (SUGEF[1]

[1] Superintendent General of Financial Institutions. It is a public entity that supervises the stability, resilience, and efficiency of the Costa Rican financial system, and inspects and regulates the operations and activities of financial entities.

M. Botto-Tobar et al. (Eds.): *Latest Advances in Electrical Engineering, and Electronics*, LNEE 933, pp. 185–199, 2022.
https://doi.org/10.1007/978-3-031-08942-8_14

14–17), which establishes within its management framework the information security processes and the critical operation of the technologies that support the business needs.

The second corresponds to the Personal Data Protection Law (8968) created to guarantee all individuals' fundamental right to informational self-determination and the processing of their data and that of their property. It requires organizations to implement actions to ensure the confidentiality of information, as well as the adoption of security measures.

The third corresponds to the Workers' Protection Law (7983), which exhorts entities to identify the institutional risk, as well as the adoption of continuous methods to preserve the proper administration of workers' resources. In this context, FONDOSP S.A., being a public financial institution that manages pension funds obtained from the savings of its affiliates representing the Costa Rican labor force, has not identified how its information security and technological failure modes, which are currently documented in its institutional risk profile, could lead to regulatory noncompliance. This problem generates two questions which need to be solved: first, what is the specific information security regulation that should be considered when assessing risks? and second, what mitigation activities could be implemented in the organization to mitigate them? In order to address the first question, we carried out a detailed investigation of the Costa Rican regulations associated with information security. For the second question, the ISO 27002 practices guide is used, prior to an analysis of the applicability status in FONDOSP recommended in the ISO27001 standard. The final proposal of this document details the control processes of applicable information security by category, providing as inputs a schematic description of the control, periodicity, type of execution and a flowchart of the parts that make up the mitigation activities.

2 Methodology

The study methodology is composed of four sequential phases: a) obtain data about FONDOSP's current situation in terms of information security, b) establish the study variables related to the field of information security, c) prepare an analysis tool for the approach of mitigation practices, and d) process the data and analyze the results obtained.

In Phase 1, an investigation based on field observation and documentary inspection was applied, for which internal and external audit reports were reviewed to obtain information related to the organizational environment, procedures, internal and external regulations supported by the organizational document management system, and technical reports executed by FONDOSP's risk and quality management department (DGRC). The results obtained at this stage are information security risks, specific regulations, and applicable organizational requirements.

In Phase 2, the information security risks previously found are mapped according to the control domains established in the ISO 27002 practice guide. As a result of this stage, the domain surveyed is justified as applicable to the object of study by establishing an inclusion or exclusion criterion. This is summarized in Table 1.

In Phase 3, an information security controls questionnaire is designed based on the ISO 27002: 2013 controls (Annex A - properly elaborated for the applicable domains that resulted from Phase 2). With this, interviews were conducted with the head of risk management in two sessions, and one session with the information security officer. As a result of the analysis, we found a gap that must first be addressed to subsequently propose mitigation activities. The tool used for the interviews is displayed in Table 2.

Table 1. Mapping of regulatory norms regarding information security control domains.

Regulatory requirement	Information security risks encountered	Applicability of the ISO 27002 control domain
SUGEF 14–17. Implementation of the Manage Security Services process (DSS05) Establish an inventory of sensitive documents and output devices and perform regular reconciliations (DSS05.06)	mismanagement of sensitive assets Findings: Relevant affiliate data collected at the end-user workstations of IT employees was detected, as well as the transfer of this data to commonly used folders	A8. Asset Management

Table 2. Interview script

Information security controls	Mitigation practices content	Support to the answers provided
A.8.Asset management	1. ¿Is the classification of information assets driven by legal or contractual obligations?	Head of risk management and quality
	2. ¿Do information assets have a risk owner and a line manager?	Information Security Officer
	3. ¿Are there review policies, standards, procedures, guidelines, and records related to the classification of information?	Institutional documentation system: quality manual, information security policies and procedures

In Phase 4, data processing is carried out in which, based on the analysis of the previous stage, the new practices required by the organization to mitigate information security risks are defined. As a result of this stage, Table 3, which corresponds to the domain "Asset Management", is outlined as an example.

Table 3. Proposed information security practices. Asset management

Control Description	Information security officer, together with the head of compliance and asset custodians, identifies information assets according to FONDOSP's processes, classifies them according to defined categories, and labels them by type and criticality. **Classification of data according to regulatory compliance:** Public, Restricted, Internal, Confidential
Frequency	twice-yearly
Responsible	Functional Unit Manager / Information Custodian / Information Security Officer
Control Type	Preventive
Evidence of implementation	Corporate inventory of information assets
Diagram of control activities	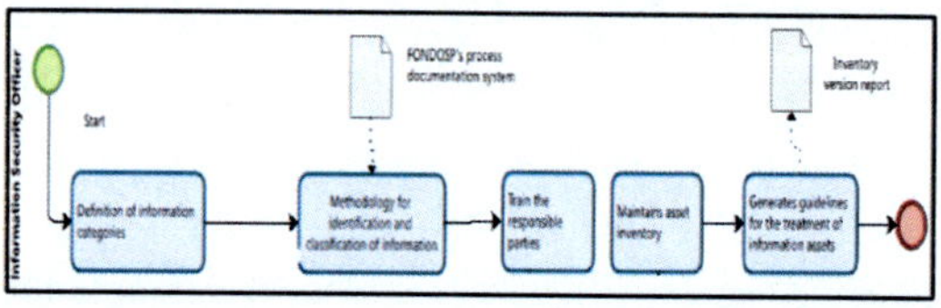

3 Results and Discussion

As a result of phase 1 of the desk research, three regulations directly applicable to FONDOSP were found and are discussed below: The Personal Data Protection Law (8968) (LPDP) [2] issued on September 5, 2011, aims to guarantee to any person, the right to privacy of their personal data, and anything related to their private life or activity. The state agency created to ensure full compliance with the provisions of this law is the Agency for the Protection of Personal Data (PRODHAB), ascribed to the Ministry of Justice and Peace. According to Sect. 1 of Sect. 2 of the law, some principles are covered such as: informed consent, confidentiality, rectification, and portability of the inhabitants' data. Its relationship with FONDOSP is specifically related to the regulation of the information stored in automated or manual databases associated with marketing services within the national territory. In this sense, the financial institution performs digital marketing and strategic advertising activities supported by internal business intelligence processes. Therefore, any person whose rights to informational self-determination or informed consent are threatened may report it to PRODHAB. Section 4 of the regulation compels organizations to make a detailed description of the type of personal data processed or stored, creating and maintaining an updated inventory of the technological infrastructure involving hardware and software and their licenses, as well as the establishment of security measures applicable to personal data, which must be effectively implemented.

Furthermore, the SUGEF 14 -17 General Regulation of Information Technology Management (RGTI) agreement, dated April 17, 2017 [3], aims to establish the minimum requirements for information technology service management by which supervised and regulated entities of the Costa Rican financial system must abide. It indicates the responsibility of risk management by the supervised entities, ensuring the information systems meet the requirements of the managed funds and interested parties. It also requires effective data management to ensure the integrity, quality, timeliness, and availability of information. This regulation is applicable since according to the technological profile presented by FONDOSP, in the System for Data Capture, Verification and Upload (SICVECA) carried out on August 14, 2018, the process related to information security called "manage security services" is declared, which is based on the COBIT 5 framework (DSS05). The purpose of this process is to minimize the impact of information security vulnerabilities and incidents on the business.

Finally, the Law for the Protection of Workers 7983 (LPT), which is of public order and social interest, was created with the purpose of regulating the funds belonging to the citizens and establishing a control system to guarantee their correct administration. One of the objectives of this law is that the workers receive a pension according to their acquired rights. The entity responsible for complying with this regulation is the Superintendence of Pensions (SUPEN), whose role is framed in the supervision and evaluation of operational, technological, financial and market risks to guarantee a correct administration of funds. FONDOSP administers three types of funds to its affiliates and beneficiaries previously declared by the Costa Rican state: mandatory pension system (ROP), the labor capitalization fund (FCL), and voluntary pension funds (FVP). These must be regulated through an adequate administration according to the law, so it is directly applicable.

3.1 Information Security Management in FONDOSP

The organization implemented an enterprise risk management methodology based on the Australian standard AS/NZS 4360:1999[2] [4], in which it integrates information technology as a component of its enterprise risk profile. In addition, the scope of this methodology encompasses cybersecurity as shown in Fig. 1.

- Inventory of technology management processes: It is based on tracking of technological production services management activities, technology vendor management, and software engineering.
- Identification of operational and technological risks: The activity consists of holding sessions with the heads of each functional unit to identify risk events, outlining them in a calculation matrix[3] [5], which contains the description of the failure mode, process, sub-process, and affected functional unit.

[2] A methodological framework widely used in the public and private sectors to guide strategic, operational and other forms of risk management.

[3] A risk matrix is a table that has several categories of "probability," "likelihood," or "frequency" for its rows (or columns) and several categories of "severity," "impact," or "consequences" for its columns (or rows, respectively). It associates a recommended level of risk, urgency, priority, or management action with each row-column pair.

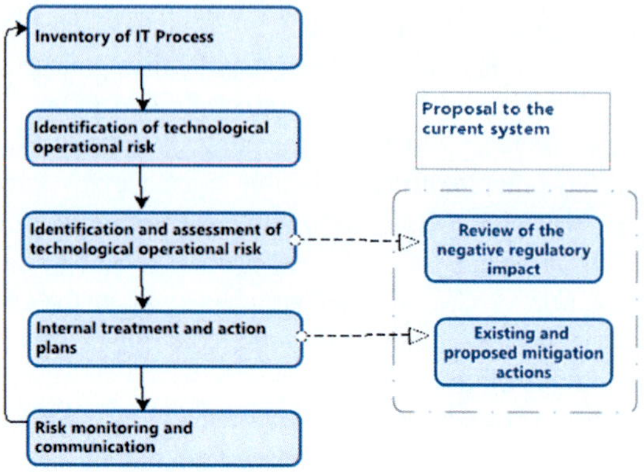

Fig. 1. Operational and technological risk management

- Measurement and evaluation of technological operational risk: According to the expert knowledge of the head of technological production services, together with the risk manager, the risks are rated according to probability and impact. The probability or frequency of occurrence is measured over a four-year horizon, and a technological event can be classified as exceptional, sporadic, moderate, frequent, or constant. Impact is measured according to four criteria: a) reputation: If, because of errors or technology failures, the image of the institution is affected, causing the general management or the mass media to become aware of it, b) clients: If, because of errors or technology failures, the interests, rights or benefits of affiliates are affected, be it slightly or severely, c) economic: If, because of errors or technology failures, losses or expenses for the organization are incurred. Even civil lawsuits may be considered. As shown in Fig. 1, it is proposed to include to the current risk assessment, a fourth factor called d) Legal: If, due to errors or technology failures, legal or regulatory events occur, a negative impact is foreseen, ranging from a written warning or regulatory fines to the possibility of suspension of authorizations to operate or to continue providing critical services.
- Plans of action: The events with high risk have a higher priority for treatment and mitigation; plans designed to reduce risk to an acceptable level should be developed. The IT manager is responsible for executing the plans and the risk manager is responsible for validating the effectiveness of the plan. The system for recording findings is used for follow-up and compliance with the plans.
- Risk monitoring: The institution monitors the indicators that compose its profile, including anti-malware system updates, availability of the corporate database and telecommunications equipment, management of roles and access to applications, and the availability of the corporate database and telecommunications equipment.

It is important to note that the methodology is focused only on technological risk management, so its indicators are focused on measuring the availability of the technological infrastructure, and not on the availability, confidentiality, and integrity of the information as a transversal asset of the organization. Another striking aspect is that there does not exist a legal risk assessment stage that considers the fines established by local regulations, such as written warnings to the organization by the supervisory agencies, and the possibility of suspension of authorizations to operate or continue providing critical services.

It became evident that identification of security risks and periodic tasks are carried out as part of the risk management department's work plan. These are organized as follows: a) corporate database monitoring. It is based on periodic reviews of critical actions executed by privileged users on the core system storage structures, b) vulnerability analysis. It is based on the discovery and correction of weaknesses in web servers, storage, and the organization's critical communication equipment, c) review of roles to critical applications. It is based on the maintenance of roles of the users of core applications, their elimination, their granting based on the segregation of functions according to the business units, d) review of roles to critical applications. It is based on the maintenance of roles of the users of core applications, their elimination, and the granting of privileges according to the users' functions of each functional unit.

As part of the mitigation of security risks, the IT Management performs compliance verification of antivirus updates. They try to keep end-user antivirus applications up to date and treat those threats detected as high risk. In addition, the risk management department conducts training on information security issues, which is based on raising awareness among the organization's employees about threats in cyberspace and how to avoid them.

3.2 Insights of the Regulatory Exposure Analysis at FONDOSP in Terms of Information Security

A set of information security failure modes was found based on the analysis of the organizational risk profile, knowledge of critical processes and their internal context. The inputs for the documentary research were technical reports prepared by the risk management department, internal and external audit findings. For the security events encountered, an analysis was performed of the economic impact dictated by local regulations.

- Senior Management does not assign responsible parties to take corrective measures to address the findings of logical security of databases or vulnerabilities of critical storage servers. There is no RACI (Responsible, Accountable, Consulted, Informed) responsibility matrix assigned within the organization for the measures. For example, we found that an IT department employee has a role that is not proper to the nature of his functions, in which he could consult personal information of clients and even balances of their financial accounts. This role corresponds to officials of the individual account management department. Thus, there is a lack of segregation of duties in the use of the corporate database.

  Regulatory Impact. If, due to lack of follow-up and application of corrective measures, personal information of members is leaked, the LPDP applies: Articles 10, 11, and 14 which correspond to a regulatory fine ranging from five to thirty base salaries corresponding to judicial assistant I (Referential amount $20,275 dollars).
- The heads of the functional units have not formally identified and classified the information assets, nor is there an owner who must identify the threats to which they are exposed. For example, the internal audit of information systems detected files

with customer contact and financial data collected at the workstations of IT management employees. In addition, the transfer of these files through common folders was also detected. This revealed the lack of care of these assets by the employees of the commercial department.

Regulatory impact. RGTI, Article 8, IT management framework processes, indicates the duty to establish an inventory of sensitive documents and output devices, and to perform regular reconciliations of these (Process "Managed security services" - DSS05.06).

Furthermore, LPDP, article 36, indicates that organizations must have an inventory of customers' and employees' personal data, as well as of their technological infrastructure and software licenses.

There are weaknesses in the specialized log management of the entity's intrusion prevention system (IPS) and firewall. It is not known for sure whether a computer connected to the internal network suffers a real attack or if it is a false positive.

- Furthermore, the technology department does not give timely attention to vulnerabilities, for example, it kept storage servers with technological obsolescence for more than a year.

  Regulatory impact. If they were to experience theft or kidnapping of personal customer information, the LPDP would apply: Article 10, 11, and 14 which corresponds to a regulatory fine ranging from five to thirty base salaries corresponding to judicial assistant I (Referential amount $20,275 dollars).
- There is no baseline for change management of the organization's critical platform, nor is there a technical analysis of the impact on the business prior to the changes. For example, the organization's critical services were unavailable on two occasions for at least 30 min due to an attempt to perform maintenance on the virtual storage of a local server.

  Regulatory impact. If, due to a change in the configuration, the critical processes of calculating quota value[4] and accounting records in customer accounts are suspended, there could be a regulatory impact for not complying with the daily report requests as established by SUPEN [6]: LPT, general provisions on the organization's requirements, Sect. 2 (Regulatory referential fine: $30,715 dollars) (Facts: failure to submit information to SUPEN).

  On the other hand, there is the RGTI Process "Manage security services" - DSS05.06 (Protect against malicious software).
- Weaknesses in logical access include a) Lack of effectiveness in the execution of logical access policies configured in data loss prevention (DLP) equipment. For example, we found that a user can connect a mobile device to the USB port of a computer and access its data. b) Unauthorized access to programming code sources of applications in production by collaborators who do not belong to the systems development department. c) Access to external e-mail and unrestricted Internet profiles where relevant information could be sent outside the organization.

  Regulatory impact. If sensitive customer information is leaked either by removable storage media or cloud, and this is used illegally, the LPDP, articles 10, 11, and 14 would apply, which correspond to a regulatory fine ranging from five to thirty base salaries corresponding to judicial assistant I (Referential amount $20,275 dollars).

[4] Quota value: minimum daily amount credited to customers as part of their savings.

On the other hand, if the leaked information corresponds to investment assets that have not been disclosed in the stock market, it is exposed to a regulatory fine of one to six base salaries, public reprimands, and even removal from management positions (Article 64 of the LPT).

Finally, RGTI, Article 8, IT management framework processes, indicates the duty to manage user security and logical access (end-user workstations, management of sensitive documents and output devices) (Process "Manage security services" - DSS05.06).

- No responsibilities and procedures have been established for the reporting, evaluation, and treatment of materialized information due to security incidents. It is evident that the organization does not maintain a database that contemplates the timely documentation of security events. For instance, in the following events the root cause was analyzed but the corrective and preventive measures or lessons learned were not documented, together with the responsible and affected parties. There was also no mechanism for classifying these incidents according to their criticality or possible economic loss. We detected: a) Loss of data from storage repositories, shared folders that served as support for the process of managing individual customer accounts, corresponding to a period of one month. This incident was not documented.

  b) Disclosure of contact and financial data of a highly reputable customer, where an employee of the commercial department mistakenly sent a mass mailing to all personnel of the organization including personal and transactional data.

  Regulatory impact. RGTI, article 8, IT management framework processes, indicates the responsibility to monitor the infrastructure to detect security-related events where the nature and characteristics of potential security-related incidents must be defined and communicated.

Both the risks identified, and the corresponding regulatory impact are outlined in Table 4.

3.3 Mapping of Mitigation Practices Based on ISO 27002

An analysis of the fourteen information security control domains associated with the ISO/NTE 27002:2013 standard was performed and mitigation practices for the identified security risks were obtained. This standard was selected because both ISO 27002 and COBIT 5 were used by SUPEN as a reference for the establishment of the information technology management regulation.

Information Security Policies

The definition of policies and procedures does not cover identified risks. For example, it is necessary to establish expected guidelines for teleworking employees and how they should manage their mobile devices.

The definition of policies should be cross-referenced with best practices such as ISO 27001 where responsibilities, scope, roles of senior management, and versioning are clearly established. The compliance officer should provide the inputs for ensuring that the regulatory criteria are met, and compliance standards are adhered to. Such is the case of the collection process of individual accounts using credit cards to collect customer

Table 4. Summary of four types of risks identified and their regulatory impact

Disclosure of confidential and governmental data	Disruption of critical services and loss of information	Inefficient management of user roles in systems	Digital abduction of valuable business data
LPDP-Art. 11,12, & 14 estimated maximum exposure: $20,275	**LPT- Art. 42** estimated maximum exposure: $30,715	**LPDP-Art. 11,12, & 14** estimated maximum exposure: $20,275	**LPDP-Art. 11,12, & 14** estimated maximum exposure: $20,275
LPT- Art. 64 estimated maximum exposure: $4,012	Business continuity impact: Loss of confidence for every 4 h of unavailable service	**RGTI – Art. 8.** Estimated exposure as determined by the superintendency	**LPT- Art. 64** estimated maximum expo-sure: $4,012 **LPT- Art. 42** estimated maximum exposure: $30,715
Exposed Assets: Personal and customer contact information, credit card information, balances, funds and investments, and value titles	**RGTI – Art. 8.** Estimated exposure as determined by the superintendency	Exposed Assets: corporate database, balances, daily quota value, accounting batch files for payment of suppliers and payroll	Business continuity impact: Loss of confidence for every 4 h of unavailable service
	Exposed Assets: corporate database, daily quota value, balances per client, daily monetary yield, account statements per managed fund		**RGTI – Art. 8.** Estimated exposure as determined by the superintendency
			Maximum total exposure: $30,715 + Loss of confidence (profitability) / 4 h of unavailable service + $20,275 + $4,012 Relevant assets: corporate database, daily value calculation, balances, statements of account by fund, contact information, securities, personal information

contributions, which are stored in the organization's internal servers. A data access security policy that considers the stipulations of the PCI DSS standard is necessary. Senior management should ensure that these and other information security policies are understood and formally accepted by employees and external suppliers. The responsibility for communicating them within the organization should not be solely that of the information security officer.

Organization of Information Security

The senior management committee should set up a committee specialized in information security which should hold bimonthly meetings to address the treatment of identified risks, make decisions on processes, action plans, and accesses to sensitive information by functional units for developing initiatives that manage data.

On the other hand, through an administrative representative, periodical contact should be established with special interest groups, professional forums and specialized security centers, where data on emerging threats, early alert warnings, and recently discovered vulnerabilities, as well as the latest cybersecurity releases, can be shared.

A matrix of responsibilities should be constructed to identify tasks in information security risk management, establishing that the first line of defense are the functional units, which must control access to critical information systems, through proper evaluation and approval of changes. A detailed list of regulatory authorities and organizations that could be contacted in case of specialized consultations, incidents, and emergencies, such as the CSIRT-CR[5] should also be established.

It is necessary to identify responsibilities for mitigating information security risks in strategic projects, system improvements, and existing processes. During the conceptualization and formalization of initiatives, the project's impact must be considered. A set of security requirements that must be covered in the different stages of project approval or deliverables should be built.

Asset Management

Establish a methodology for the classification of information assets supported by an analysis of contractual, legal, or regulatory obligations, which allows the establishment of asset categories according to their criticality.

Build a formal inventory of assets that involves the participation of each functional unit, and contemplate digital data, printed information, software, infrastructure, service providers, physical devices, and people.

Design an internal guideline that considers directives on acceptable use of technological assets, personnel that must use it, or the use of valid personal mobile devices to process sensitive information for the company.

Establish a baseline for the classification of information according to its level (conf dential, restricted, internal, and public) as well as their access rights over it (read, write, and delete).

Perform labeling and assignment of access permissions to assets by personnel, using watermarks or seals according to previously established nomenclatures.

---

[5] Costa Rica Computer Security Incident Response Center (CSIRT-CR), part of the Ministry of Science, Technology and Telecommunications (MICITT).

Establish directives for disposal of storage media or media to ensure that particularly sensitive data is disposed of securely (cryptographic erasure, degaussing or physical destruction). It has been identified that personnel use paper shredding machines for physical destruction of information (Measures influencing compliance with LPDP, articles 29, 30, 31).

Security of Operations and Communications
Incorporate a SOC team to the production and technological infrastructure department that ensures the execution of anti-malware detection and correction policies for all end-user equipment, including mobile devices. In case of detected events, proceed with the escalation and treatment of IT security events.

Implement an information security event management system (SIEM) to generate indicators of compromise (IoC) related to changes in user IDs, privileged system activities, successful and unsuccessful access attempts, software installation, accessed files and types of access. The above for the purpose of acting on information security incidents.

Every time software is installed or developed, it needs to be tested in a segregated and secure environment, in which a risk analysis is performed, and sent for approval to the functional unit that authorizes the use of the software.

Permanently execute computer security audits to measure the effectiveness of access policies to the organization's critical systems, access to external storage devices, anti-malware detection and prevention. A flow of activities is proposed as shown in Fig. 2.

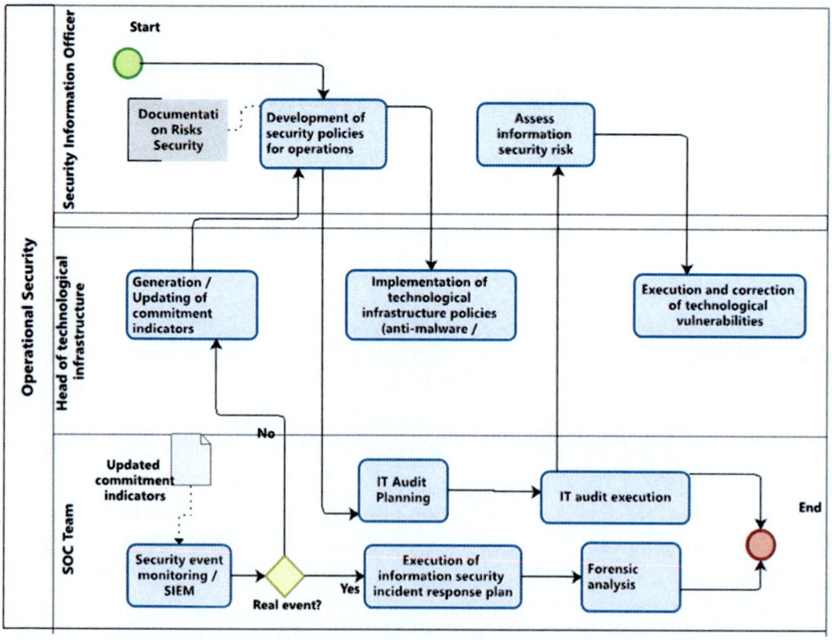

Fig. 2. Proposed flowchart on safety of operations

Access Control
An information security risk committee should conduct periodic reviews of the type of critical asset (confidential, internal, or public) accessed. Indicators of compromise must be implemented on network access connections, taken from weekly logs of IPS equipment or others, to respond to positive alerts, or implement additional controls such as Dual Factor Authentication for access to information or one-time numeric codes (OTP).

Implement a procedure with the human development and organizational health department to register and deregister users from the systems, as well as update the inventory of access to information systems.

Conduct a review of privileged accounts for access to critical information systems, source code storage units, or versioning repositories.

Information Security Incident Management
Assign a functional unit expert to perform the root cause analysis of an information security incident, classify it and propose corrective and preventive measures. Establish guidelines to evaluate incidents, their recurrence and perform escalation to stakeholders according to their criticality. For high criticality events, a crisis or information security risk committee should be established.

In conjunction with the information security officer and expert users, they should collect evidence of incidents and maintain a chain of custody in the storage and transfer of interested parties.

Establish a maximum time frame for incident reporting and documentation. For instance, it can be done during the first five days after the event is discovered, depending on its criticality.

Consider in the incident management methodology that any incident that has resulted in a material economic loss must be reported in the accounts assigned for this purpose.

Note that any incident that has complied with the preventive measures or its action plan must be closed once the lessons learned have been communicated to the stakeholders.

4 Conclusion

We propose eliminating subjectivity in the assessment of FONDOSP's own information security risks by specifying the regulatory impact in detail, to estimate the expected value in the materialization of failure modes; in addition to providing guidelines to know which specific aspects of the regulations must be considered.

As a result of the research, the risk associated with abduction or disclosure of critical information, which contemplates both the interruption of services and the loss of sensitive information, may be regulated by Law 8968 - protection of personal data; Law 7983 - worker protection; and the General Regulations for the management of information technologies for the financial sector. The maximum economic exposure for non-compliance with these regulations together could exceed $55,000, in addition to the loss of business profitability for every 4 h of unavailable customer service.

The mitigation practices that the organization is currently suffering from are not only related to the design of technical controls applied to the technological infrastructure, but also to the leadership, responsibility, planning and support to the processes by the top management regarding information security management. We determined that the technological operational risk management methodology does not consider the negative regulatory impact.

Among the main practices of organizational security is the need for an information security risk committee in charge of making decisions when authorizing or denying changes, implementing corrective and preventive measures to incidents, determining critical accesses and, above all, establishing procedures to protect the value of the assets.

A very necessary mitigation practice is the management of information assets in which joint responsibility activities are proposed between the expert collaborators of the functional units, the information security officer and the compliance officer to carry out an inventory of assets according to a methodology that allows categorizing, identifying and classifying information, training personnel, and establishing guidelines for the treatment and use of assets.

Some important practices in operations management have to do with the incorporation of a Security Event Monitoring System (SIEM), which would allow adopting a proactive approach in the detection and correction of information security incidents based on the incorporation of key Indicators of Compromise (IoC). This would allow prioritizing physical and economic resources to protect the organization from potential threats and not just from false positives. We propose the integration of a SOC team to carry out these operational security tasks.

In order to guarantee the confidentiality of information assets regulated by the personal data protection law, such as names, salary, marital status, beneficiaries, telephone numbers, e-mails, credit and debit card numbers, we propose strengthening security measures related to access to corporate core support media and applications, such as the implementation of Double Factor Authentication (DFA), the use of One-time Numeric Codes (OTP), Data Loss Prevention (DLP) for e-mail and removable media. This is even more necessary to prevent the loss of sensitive information when personnel are teleworking.

On the other hand, some information assets associated with the worker protection law include balances in individual accounts of affiliates, investment securities, daily quota value and accounting batching macros. These must be protected, not only for regulatory reasons, but also because of the criticality they represent for daily operations.

References

1. Barahona, C., Zamora, D.: Valuation of the digital public experience. INCAE Business School, Alajuela Costa Rica (2019)
2. Attorney General's Office (PGR) Law 8968: Homepage. last accessed 08 Oct 2020
3. Superintendent of Pensions of Costa Rica (SUPEN): General Technology Management Regulation. Costa Rica (2011)
4. Standards Australia: AS/NZS 4360–1999 Homepage. https://www.standards.org.au/standards-catalogue/sa-snz/publicsafety/ob-007/as-slash-nzs--4360-1999. last accessed 05 June 2020
5. Cox, L.: What's wrong with risk matrices? In: Society for Risk Analysis, Risk Analysis, vol. 28(2), p. 497. Denver, United States (2008)

6. Superintendent of Pensions of Costa Rica (SUPEN): Information Manual for Supervised Entities and Managed Funds. San Jose, Costa Rica (2020). Homepage, https://www.supen.fi.cr/documents/10179/148522/Manual+de+Informaci%C3%B3n+para+las+entidades. last accessed 20 Oct 2020
7. INTE ISO IEC 27001:2014: Information technology — Security techniques — Information security management systems — Overview and vocabulary. San José, Costa Rica (2018)
8. NIST SP 800–30: Information Security. Guide for Conducting Risk Assessments. Revision 1. U.S Department of Commerce, Gaithersburg, United States (2012)
9. Superintendent of Pensions of Costa Rica (SUPEN): Supervision Framework, https://www.supen.fi.cr/marco-de-supervisión
10. Valencia, F., Marulanda, C.: Governance and management of information technology risks and aspects that differentiate it from organizational risk. National University of Colombia. Bogotá, Colombia (2016)

Telecommunications

Analysis and Simulation of Downlink Scheduling Algorithms on 5G NSA Networks Under FTP Traffic

Javier Márquez, Pablo Lupera Morillo(✉), and Luis F. Urquiza-Aguiar

Departamento de Electrónica, Telecomunicaciones y Redes, Escuela Politecnica Nacional, Quito, Ecuador
{javier.marquez,pablo.lupera,luis.urquiza}@epn.edu.ec

Abstract. This research presents the performance of Round Robin, Proportional Fair and Maximum Throughput scheduling techniques for 5G NSA mobile networks. The tests run considering a variable and increasing number of users using the FTP Model 3 traffic scheme, and a low frequency band (1800 MHz FR-1) to know how the throughput, delay and spectral efficiency behave under these circumstances. In general, the results show that either Proportional Fair or Maximum Throughput schedulers can be a good choice for FTP traffic. Even though Round Robin provides a throughput as high as the one obtained by the Maximum Throughput scheduler (15% less), the delay introduced to all users is also high. For ITU requirements on eMBB services, we found that only the delay at User Plane level could be satisfied.

Keywords: Scheduling algorithms · 5G · NSA · MATLAB

1 Introduction

The path of evolution from 4G to 5G, specially for carriers and mobile operators is about planning and investment in all the new infrastructure, this is to provide a full end to end solution so that new services' requirements can be fully met.

First, the vision of the ITU-R organism about the services that 5G should provide/support is shown in the next figure:

M. Botto-Tobar et al. (Eds.): *Latest Advances in Electrical Engineering, and Electronics*, LNEE 933, pp. 203–218, 2022.
https://doi.org/10.1007/978-3-031-08942-8_15

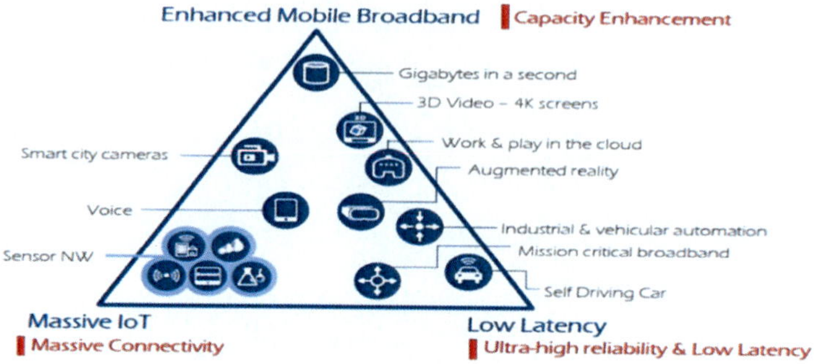

Fig. 1. 5G services/requirements [1]

In Fig. 1, the three main "new" services or use cases supported by 5G are:

- **eMBB** (Enhanced Mobile BroadBand) applications that require a really high throughput. Requirement: 20 Gbps in downlink
- **URLLC** (Ultra Reliable Low Latency) applications that require an extremely low latency and high reliability during transmission Requirement: user plane delay about 1 ms for small packets
- **mMTC** (Massive Machine Type Communications) scenarios that require to support many connections simultaneously Requirement: 1 million connections per Km^2

One important aspect about the access network is the fact that the frequency spectrum at low frequency range is either scarce or rather congested. This issue brings one of the main features of 5G networks, the use of a much wider spectrum, from a low frequency range (the same used by GSM/3G/4G) up to mmWave range (see Fig. 2).

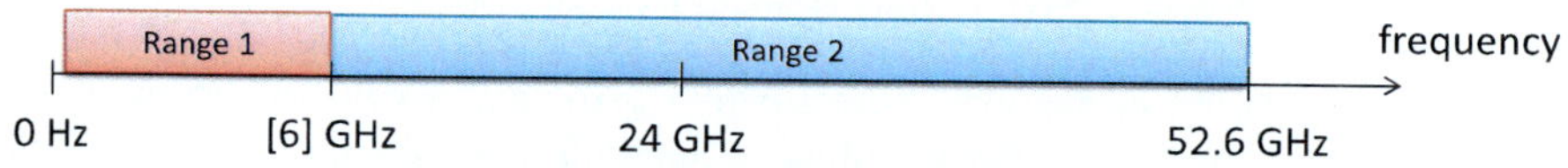

Fig. 2. 5G spectrum division [2]

Figure 2 evidences the increase in the spectrum assigned to 5G. Many researches have been performed focused mainly in the mmWave spectrum (formally named FR-2) because of the trade-off that needs to be accomplished between the really harsh channel condition that occurs when the frequency increases and the capacity of the channel (up to 400 MHz so far) that is available to use. If on FR-2 there is much spectrum to exploit, what occurs at low frequency? In the FR-1 band placed below the 6 GHz, we found the opposite: the channel conditions are better to transmit but the spectrum is so populated that frequency planning should be done with care.

This is why the scheduling techniques are important at low frequency (the fact that at high frequency there are plenty resources, does not mean schedulers are not important to use as well). The scheduling technique decides how to allocate resources according to the requirements of transmission of each user, a critical process especially at sub-3GHz band where the channel bandwidth may not be sufficient for some 5G new services.

In order to speed up the migration pace to 5G around the globe, the 3GPP standardization team divided the deployment of 5G in two phases [3]. PHASE I introduces 5G SA (Stand Alone) and NSA deployments and they are detailed in Release 15 documentation. The NSA architecture aims to support the early deployment of 5G services. In this work, we carry on a performance evaluation of Round Robin, Proportional Fair, and Maximum Throughput scheduling techniques for 5G NSA mobile networks.

The rest of this paper is organized as follows: Sect. 2 introduces background on schedulers and data traffic generation. Then, Sect. 3 presents the related work. Next, Sect. 4 offers details about the research method that we use. After that, Sect. 5 analyses our test results Finally, Sect. 6 presents the conclusions obtained with this work.

2 Background

The traffic for all active users should be scheduled considering the queue state and the channel conditions per user. For this reason, Round Robin, Proportional Fair, and Maximum Throughput are considered basic scheduling algorithms [7] where the throughput is the most crucial factor. The delay and spectral efficiency are derivate parameters from throughput measurements. These parameters are very important on eMBB scenarios. In this section, we summarize some key ideas about these schedulers. In addition, we describe the data traffic model that we used in our tests.

2.1 Scheduling Techniques

OFDMA divides air resources in time and frequency resources. In frequency, the channel bandwidth is divided in RBs (Resource Blocks). Every RB bundles 12 subcarriers separated by subcarrier spacing or SCS. On 5G NSA networks, as there is a co-existence between LTE and NR nodes, only 15 kHz SCS is supported, giving a transmit time interval TTI of 1ms. In time domain, the frame duration is 10 ms, each frame has 10 sub-frames of 1ms, and on each sub-frame a number of timeslots of duration TTI is transmitted (for SCS = 15 kHz, 1 slot per sub-frame is used).

Round Robin (RB): This algorithm does not consider the channel state of the users. Instead, all resources available are shared uniformly and cyclically between them in each timeslot (meaning there's no priority).

The throughput can be calculated as [8]:

$$\text{Throughput} = \frac{\text{TBS}}{\text{TTI}} \tag{1}$$

where the TTI denotes the duration of a slot and the TBS (Transport Block Size) is the data result of the Radio Stack encapsulation process that is delivered over the air interface.

Maximum Throughput (MT): This scheduler allocates every RB resource per slot of time only to the users that achieve the highest throughput at this particular instant of time and frequency.

The priority i on the resource block K for this case is based only on the instant throughput of each user so the allocation can be modeled with the following equation [8]:

$$\mathrm{i(k) = argmax(Throughput_j)} \tag{2}$$

where the Throughput of user j can be calculated using Eq. (1).

Proportional Fair (PF): This algorithm works between RR and PF methods. The working principle is the same as in MT since the user with the highest priority is chosen, but the priority is calculated as follow [8]:

$$\mathrm{i(k) = argmax(\frac{THROUGHPUT_j}{T_j(k)})} \tag{3}$$

Again, $\mathrm{THROUGHPUT_j}$ can be calculated using (1), and Tj denotes the throughput evolution over a certain past window and it is calculated as [8]:

$$\mathrm{T_j(k+1)} = \begin{cases} \left(1-\frac{1}{t_c}\right)\mathrm{T_j(k)}+\frac{1}{t_c}\mathrm{THROUGHPUT_j(k)} \text{ if user j is assigned RB K} \\ \mathrm{T_j(k+1)}=\left(1-\frac{1}{t_c}\right)\mathrm{T_j(k)} \text{ otherwise} \end{cases} \tag{4}$$

The parameter Tj(k) adjust the priority of the users with better channel conditions so that other users with worst channel conditions can be more likely to be assigned resources and in consequence their throughput increases.

The variable tc ($1 < tc < \infty$) is the past window or number of time slots used to modify the average throughput [9]. If $tc = \infty$, then the decision is made only by using the instant throughput (like the MT scheduler). On the other side when $tc = 1$, the algorithm becomes more fair (all users receive the same quantity of resources just like RR scheduler) [10].

Once throughput results are obtained, the delay at access network level and spectral efficiency are calculated with the following equations [11]:

$$\mathbf{T_i} = \frac{\mathrm{A_i}}{\mathrm{C_i}} \tag{5}$$

$$\mathbf{SE_i} = \frac{\mathrm{C_i}}{\alpha^{(j)} * \mathrm{AB}} \tag{6}$$

where C denotes the throughput, A the size of the packet, AB the bandwidth assigned, $\alpha^{(j)}$ is a scaling factor (for FDD it has a value of 1) and i represents the user index.

2.2 Data Traffic Model

As our research is focused on 5G NSA networks we have to consider a good traffic model. Hence, we choose the FTP 3 traffic model because 3GPP classified it as intensive eMBB traffic [4]. Three main parameters define how FTP 3 traffic works: the file size S, the inter-arrival time (t_a) and the reading time (t_d). Figure 3 shows how these parameters generate FTP traffic.

Figure 3 also shows that when a file is transferred, a fixed size or a random size file is first generated and then broken into multiple packets. This is often transmitted at a constant rate, but the arrival time may not show this behavior (the inter-arrival time may not be fixed). The reading time is the delay that occurs when the receiver completes the reception of a file and sends a request to start a new transfer (delay between file transfers).

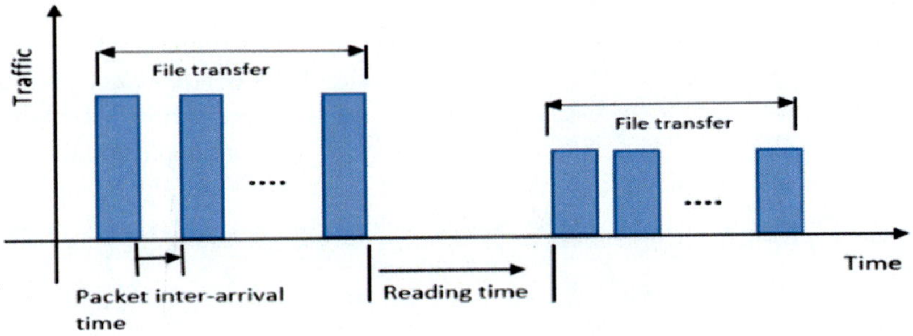

Fig. 3. FTP 3-model traffic pattern [5]

FTP traffic mathematically can be expressed using probability equations [6]:

$$\text{For file size } \mathbf{S}\text{:} \quad \text{PDF:} f(x) = \frac{1}{\sqrt{2\psi x}} e^{\frac{-(\ln(x)-\mu)^2}{2\sigma^2}}, x > 0, \sigma = 0.35, \mu = 14.45 \tag{7}$$

$$\text{For reading time } \mathbf{td}\text{: PDF}: \ f(x) = \lambda e^{-\lambda x}, x \geq 0, \lambda = 0.006 \tag{8}$$

3 Related Work

Since previous generations, the scheduling methods for wireless networks have been researched because the standard does not define the scheduling strategy. Instead, this is vendor-specific. In [12], Round Robin, Proportional Fair and Best CQI schedulers are studied for LTE, here the algorithms work by assigning available resources from the air interface to all users multiplexed according to the channel conditions of each user. This is done because in 4G, OFMDA multiplexing technology was introduced in replacement of CMDA (Code Division Multiple Access). The necessity to achieve LTE transmission requirements (for example 100 Mbps in downlink and 50 Mbps in uplink) is one of the reasons behind the importance of the scheduler existence, so that other mechanisms such us the CQI Report can provide valuable information to it and so enhance the experience of the final users.

The research goes one-step further in [14]. In [13], it is mentioned that Round Robin and Maximum Throughput are the extreme scheduling cases. For Round Robin, the channel conditions of each user is just ignored while for Maximum Throughput, the channel quality is the only parameter taken into account to assign resources. Proportional Fair works in the middle of these two last mentioned algorithms, and again the channel quality defines the priority of each user. This is why authors in [14] focused on study analytically and by simulation the effects of fading channels to the performance of the Proportional Fair algorithm. The conclusion states that when the SINR is higher (i.e. in Rician environments versus Rayleigh fading) the throughput that provides the scheduler is also higher.

In [15] a scheduler that supports both eMBB and URLLC traffic is developed by modeling math expressions with defined constrains. The objective is to maximize a utility function first, then puncture the URLLC traffic over the new flexible frame structure that conveys the eMBB data. This is a difficult approach for future applications because the design not only considers how to allocate the resources according to the defined constrains but also it needs to include in the equation the retransmission mechanism at MAC level because of the punctured data. This paper demonstrates that a proper design can be done by considering the requirements of the communication system, QoS parameters and the very flexible frame structure, leveraging all the new features introduced in 5G systems.

As our research is focused on the early stage of development and deployment of 5G, the results of [16] may be used as a reference since it uses Round Robin and Proportional Fair scheduling techniques on the mmWave band of 5G on a NSA network. The contrast with the research presented in this document is that the simulator used is MATLAB instead of the NS-3 and, the low band with simple AWGN channel is used to understand the behavior of the algorithms. This model can be used in future works assuming more limitations and conditions to make a more complete and complex scheduling solution.

4 Research Method

Simulations at access network level were performed on this research by using MATLAB R2020B and its 5G toolbox. This choice allowed us to comply with the 3GPP Rel.15 guidelines and get reliable results. We use the diagram presented in [17] on Fig. 4 as the scheme is very similar with our simulation solution.

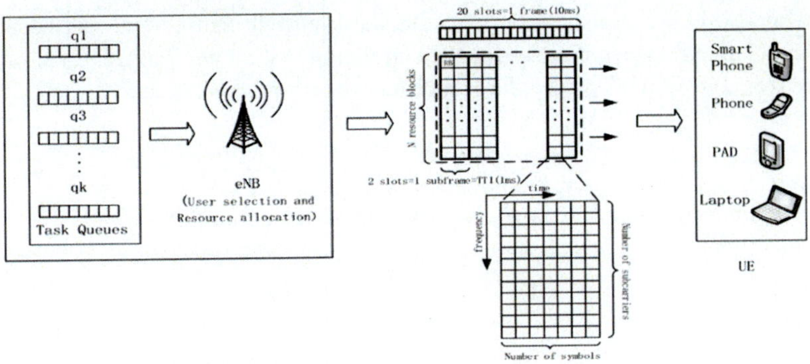

Fig. 4. 5G NSA simulation scenario [17]

The simulation process starts by generating FTP Traffic. The file size (Eq. 1) is modeled as a truncated lognormal distribution. The maximum file size S is 5 Mbytes for a tipical FTP-3 traffic source, although for intense eMBB applications, smaller sizes are possible like 0.5Mbytes or even 0.01 Mbytes [4]. To obtain the expression that generates a random file of size S, we first remember that a lognormal distribution is obtained from a continuous random variable S:

$$Y = \ln(S) \tag{9}$$

The result is that Y is another continuous random variable that has a normal distribution with mean μ and standard deviation σ [18]. Now that we see the relation between the lognormal and normal distributions, we can solve this problem by applying a linear transformation (formally called Projection to Standard Normal [19]). Here any normal random variable can be transformed by subtracting the mean and dividing by the standard deviation, in this case the RV (random variable) Y complies with this rule:

$$Z = \frac{Y - \mu}{\sigma} \tag{10}$$

As this is a very common case of use, this Z transformation is used to express the CDF (Cumulative Density Function) of Y in terms of the new variable Z [19]:

$$\text{CDF: } F_Y(y) = P(Y \leq y) \tag{11}$$

$$F_Y(y) = P\left(\frac{Y - \mu}{\sigma} \leq \frac{y - \mu}{\sigma}\right) \tag{12}$$

$$F_Y(y) = P\left(Z \leq \frac{y - \mu}{\sigma}\right) \tag{13}$$

$$F_Y(y) = \phi\left(\frac{y - \mu}{\sigma}\right) \tag{14}$$

From Eq. 10 and 14 we solve for y and we get:

$$Y = \mu + s * \phi^{-1} \tag{15}$$

On Eq. 14, ϕ denotes the CDF of a standard normal distribution, and because it results in a value between 0 and 1 for the simulation, we model it as a normal random variable from 0 to 1. The last step is to return to the original variable S:

$$\mathbf{S} = e^{\mu + \sigma * \phi^{-1}} \tag{16}$$

The reading time (Eq. 2) shows an exponential behaviour with $x > 0$, so the PDF results in:

$$y = \int_0^{t_d} \lambda e^{-\lambda t} dt \tag{17}$$

$$y = 1 - e^{-\lambda t_d} \tag{18}$$

Now solving Eq. 10 for t_d:

$$\mathbf{t_d} = -\frac{1}{\lambda} * \ln(1 - y) \tag{19}$$

Knowing that $0 \leq y \leq 1$, and from [6] $\lambda = 0.006$ s and the mean $1/\lambda$ is 180 s the reading time is then:

$$\mathbf{t_d} = -180 * \ln(1 - y) \tag{20}$$

Finally, we use a Poisson distribution process for the inter-arrival time. From [4] the mean inter-arrival time required is 50 ms, then we assume that 1 packet should arrive on average every 50 ms or 20 packets/s so the mean $= \lambda t = 20$. The Poisson process for the inter-arrival time ta is then [18]:

$$\mathbf{t_a} = \frac{e^{-\mu} * \mu^k}{k!} \tag{21}$$

Where the possible values for k are 2, 5, 8, 10, 14 according to the 3GPP technical specifications. One important annotation for t_a is that this parameter affects the queue behavior. In order to simplify the implementation in MATLAB, we assume that the inter-arrival time should be of at least 1 ms. We also assume that the buffer located at the gNodeB is divided in single queues for each active user (see Fig. 4).

This simplifies the buffer simulation complexity as on every time step of 1 TTI duration, a maximum of 1 packet can arrive to each queue. We also assume that on every time step a maximum of 1 packet per queue can leave the system, so that every queue behaves like an M/M/1/B queue [20].

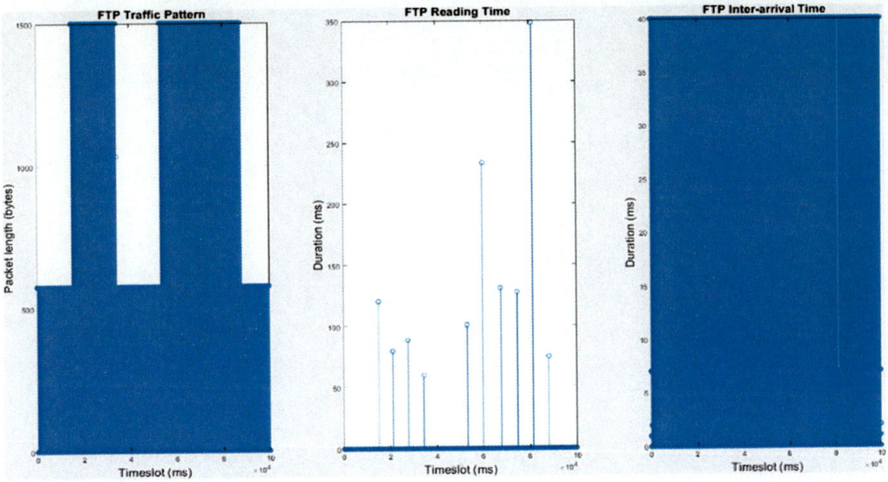

Fig. 5. FTP model 3 traffic generation

Figure 5 shows an example of FTP traffic generated using Eq. 16, Eq. 20 and Eq. 21. It is noticeable how the maximum packed size is limited by the MTU. We also got a maximum inter-arrival time of 40 ms and the most important parameter is the reading time. For example, in the 15017 ms we got a td = 121 ms. This is why in the traffic pattern shown in the left, the MTU has changed from 576 bytes to 1500 bytes. This also indicates that one file with MTU 576 bytes finished its generation and after 121 ms, another file is generated but this time with an MTU of 1500 bytes.

Once we review the FTP traffic generation process, Table 1 summarize the main simulation parameters. Figure 6 shows how the FTP generator works together with the rest of the simulation system.

Table 1. Simulation parameters

Parameter	Value
Carrier frequency	1800 MHz
Channel bandwidth	10 MHz (52 RB)
Number of gNodeB	1
User mobility	Constant position
Number of users	5,15,30,50
Transmission scheme	SISO
Traffic model	FTP 3
Buffer length	5 packets (per user)
Scheduling technique	RR, PF, MT

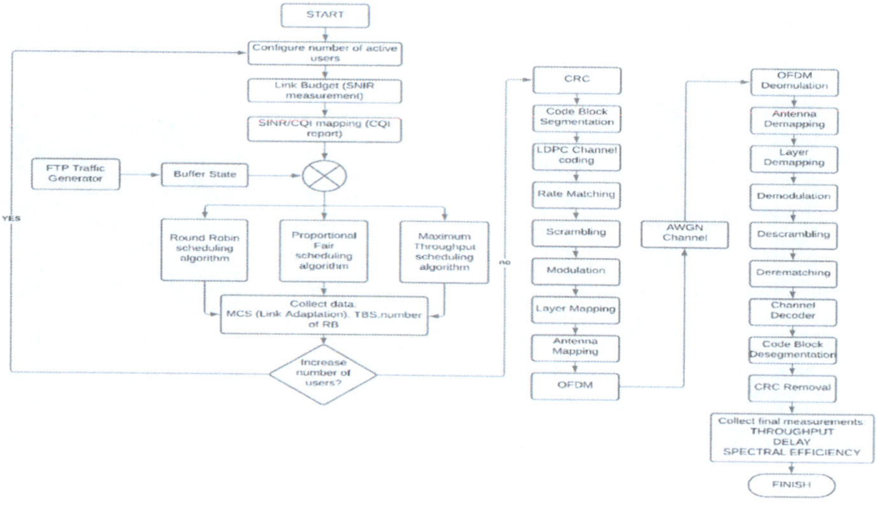

Fig. 6. Flowchart system

A simple link budget was implemented based on a 3GPP template [21] to measure the SINR of each randomly located user around the unique 5G base station. The scheduling decision is taken on each timeslot over all the available RBs according to the channel conditions of the chosen user.

The size of each packet is configured using the IP MTU (Maximum Transmission Unit) length, the FTP packet distribution stablishes that 76% of the transmissions use a 1500 bytes MTU and the rest 24% an MTU of 546 bytes [6]. Finally, using the 5G toolbox we model the DL-SCH (Downlink Shared Channel) and PDSCH (Physical Downlink Shared Channel) to transmit all generated packets using simulation parameters listed on Table 1, finally obtaining throughput, delay and spectral efficiency results.

5 Result and Analysis

A quick summary of the results obtained using MATLAB is shown next. The main focus is to measure the how throughput, delay and spectral efficiency behave when the density of users increases.

5.1 Throughput Simulation Results

The results for minimum (min) and maximum (max) throughput values obtained for the three scheduling algorithms is shown in Fig. 7. We established in Section I that the scheduler Round Robin shares all availale resources equally between all users. This form of resource allocation means that the more active users present on the cell, the lower the throughput they experience as all resources are shared equal and ciclycally. In our implementation, as we are considering M/M/1/B queues, we will simulate Round Robin by transmitting 1 packet of a user at a time [8].

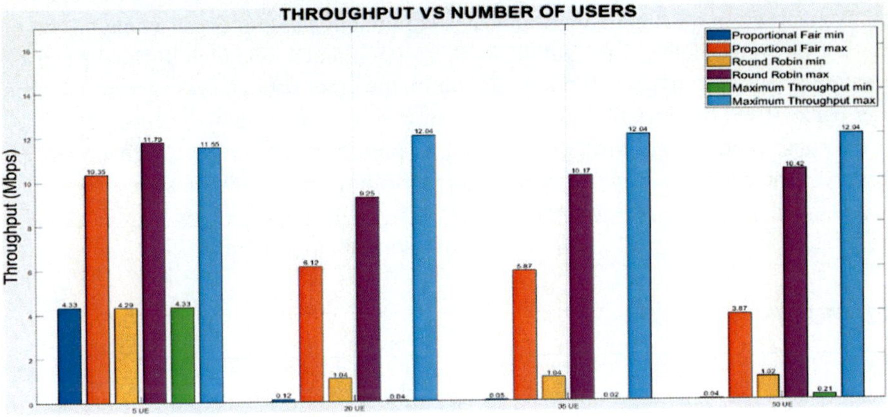

Fig. 7. Throughput vs number of users for FTP-3 traffic

This means that we ciclically transmit a single packet per user, but we allocate all resources needed to transmit that packet. In other words, only when the packet of the first user has been completely transmitted, we move to user 2 and start transmitting its packet, and so on. This is why in our implementation Round Robin can get throughput values almost as high as the Maximum Throughput approach, but serving to all users. To clarify this let´s see Fig. 8. In Fig. 8, the resource allocation representation for 5 users during 1 frame or 10 slots of time is shown. Each color associate a number, and each pair (color, number) represents a user (all except number 0 – dark blue color that represents idle or available resources).

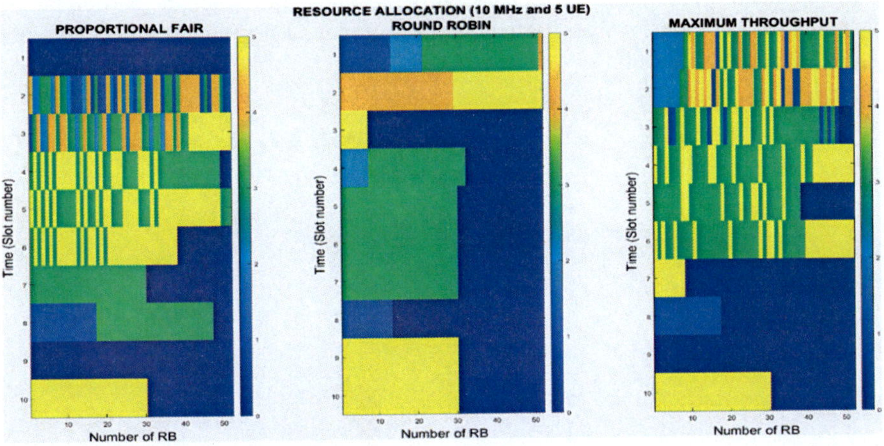

Fig. 8. Resource allocation for 5 users

We can see that Round Robin allocates continuous RBs until the whole packet of every user is completely transmitted. As many continuous RBs are allocated per user, the throughput gets higher. Proportional Fair on the other hand, works by sending each user packet little by little [22]. This allows more users to be multiplexed in frequency at the same time, so the throughput is not as high but the fairness or chance to get resources increases. Maximum Throughput algorithm serves only those users with the best channel conditions, meaning the rest of the users that do not report a good CQI will suffer starvation as they receive some or any resources all the time.

5.2 Delay Simulation Results

Figure 9 depicts the delay result when the number of randomly located users increase. When 5 UEs are being served, Proportional Fair and Maximum Throughput are able to transmit 1 packet in under 2 ms, while Round Robin takes up to 7 ms to transmit a whole packet per user. When 20 users are active, PF and MT allow that the users with the best channel conditions can still transmit a packet in about 1 ms. The worst users that report low CQI values with PF transmits in average 1 packet every 97.79 ms or 0.097s. MT introduces a delay of 272.07 ms or 0.27 s for the worst users since they receive very few resources to transmit. Round Robin provides between 48.25 ms and 58.62 ms of delay for the best and worst channel users respectively.

Finally for 50 users, the results show that the best users can still be able to transmit a single packet in about 1 ms using the PF or MT algorithms. A big delay of about 341.09 is introduced by PF because it is multiplexing many more users per slot of time, while MT keeps is value of 57.69 ms as not all 50 active users receive resources in any time. For Round Robin we have to clarify that the delay between all users is almost the same because one user has to wait one round time. When there was only 5 users, user 1 waits until the other 4 users transmit their packets to transmit again. When the base station is allocating resources to 50 users, user 1 has to wait that all the other 49 users transmit before it can transmit again.

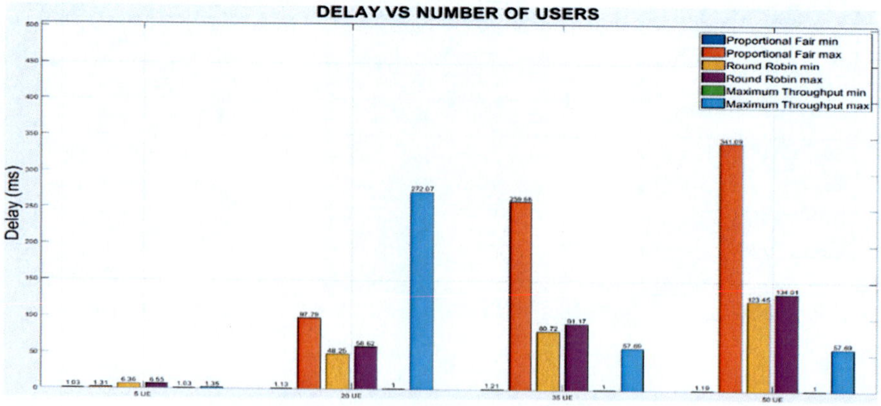

Fig. 9. Delay vs number of users for FTP-3 traffic

5.3 Spectral Efficiency Simulation Results

A summary on the spectral efficiency results for various number of users is shown in Fig. 10. The maximum values show pretty similar values for the three algorithms: 3.11 bit/s/Hz for 5 users, around 4.35 bit/s/Hz for 20 users (40% more), 5.56 bit/s/Hz for 35 users (27% more) and around 5.93 bit/s/Hz (6.65% more) when 50 users are being served. This behave may be explained by considering the Eq. 6. The spectral efficiency is proportional to the throughput of each user and inversely proportional to the bandwidth occupied to transmit.

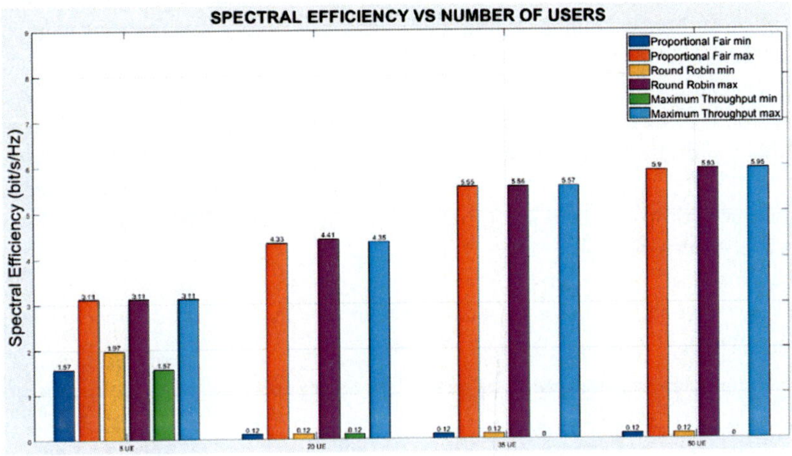

Fig. 10. Spectral efficiency results on change number of users for FTP-3 traffic

When more users are added to the cell, the throughput of the users may be compromised, as the scheduler needs to allocate more users in the Resource Grid (the representation in time and frequency of the physical channels used to transmit data). Here the results increases with the number of users because more users with better channel conditions are present. They have a better CQI so the modulation and coding scheme is also high, experiencing a higher throughput but in contrast, as there are more users, the number of RBs they could get may not be very high. The result is that their spectral efficiency increases. A similar analysis may be applied to the minimum values, but this time considering that their throughput, number of RBs per user or both decreases, so their spectral efficiency in consequence get lower.

5.4 Simulation Results Summary vs Requirements

Table 2 summarizes the main results and the requirements established by the ITU organization. We can see that only the PF or MT algorithms can satisfy the delay requirements at the user plane level with up to 50 users. Round Robin transmitting one packet at a time per user gets high throughput values, but the delay is always high. This brings another problem. For the simulation stage, we assume a queue that can store 10 packets per user, so it can overflow if packets are not dispatched quickly.

Table 2. Simulation Results Vs Requirements In Downlink

		REQUIREMENT	PROPORTIONAL FAIR		ROUND ROBIN		MAXIMUM THROUGHPUT	
			5 USERS	50 USERS	5 USERS	50 USERS	5 USERS	50 USERS
BEST RESULTS	THROUGHPUT (Mbps)	100	10,35	3,87	11,79	10,42	11,55	12,04
	DELAY (ms)	4	1,03	1,19	6,36	123,45	1,03	1
	SPECTRAL EFFICIENCY (bit/s/Hz)	7,8	3,11	5,9	3,11	5,93	3,11	5,95
MINIMUM RESULTS	THROUGHPUT (Mbps)	100	4,33	0,04	4,29	1,02	4,33	0,21
	DELAY (ms)	4	1,31	123,45	6,55	134,01	1,35	57,69
	SPECTRAL EFFICIENCY (bit/s/Hz)	7,8	1,57	0,12	1,97	0,12	1,57	0

We can see in Fig. 11 at the right-hand side, the number of packets transmitted by each scheduling technique. While the PF and MT algorithms transmit hundreds of packets, Round Robin transmits between 189 and 207 packets in presence of 5 users, but it only gets to transmit 26 or 27 packets when 50 users are present. This is because of the delay introduced by the scheduler to all users. The other issue Round Robin has is that the number of dropped packets (left-hand side in Fig. 11) is very high for all users. The reason for this is the behavior of the M/M/1/B queue: only one packet per user leaves at a time, but during the dispatch time, more packets arrive. The queue overflows and packet loss is very evident.

The PF and MT algorithms behave better as some users are able to transmit all their packets (even when the number of active users is 50), having zero dropped packets.

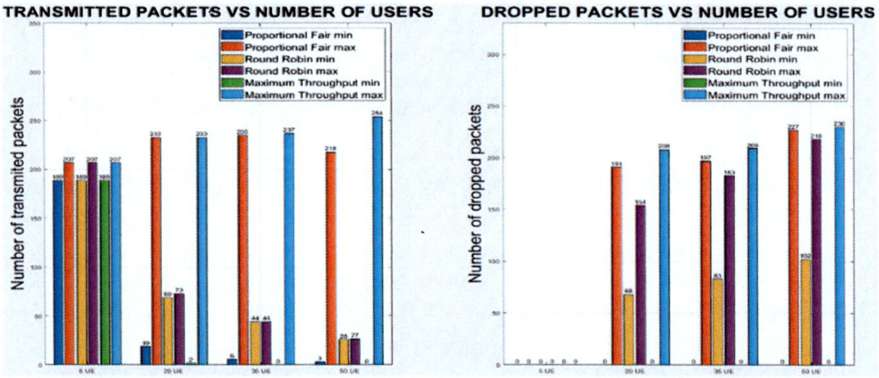

Fig. 11. Number of transmitted/dropped packets vs number of users

The throughput of 100 Mbps is more difficult to achieve, but we have to consider which applications do need this very high throughput. To the best of our knowledge, vendors like Huawei have demonstrate some eMBB applications that requires high throughput like 4K video (it requires 25 Mbps). This shows that the throughput requirement on 5G could and will be meet using other complementary technologies like multi-antenna transmissions, channels with more capacity (like the ones in the mmWave range that can use up to 400 MHz) and even higher modulation techniques.

The final parameter we have found of interest is the payload. In our research, for 1 Mbyte files transmitted within packets of 1500/576 bytes, a throughput of 12.04 Mbps was achieved when the channel conditions were the best. This indicates that we need to transmit more packets (more data) at a time to increase the throughput (if we just keep transmitting one packet of 1500 bytes to each user, we will never get that 100 Mbytes throughput experience to the users). For the spectral efficiency requirements, the same comments than that for the throughput apply. With MIMO techniques or better MCS using the same bandwidth, we can reach higher values of spectral efficiency.

Unfortunately, we did our best to find any other material related to the scheduling analysis and/or simulation process in the low frequency band of 5G, but all other works does not match with our simulation parameters. This becomes our motivation however. Even though we can compare directly our results to other works, we can clearly conclude something; the behavior of the RR, PF and MT algorithms is expected to be the same no matter the frequency of the carrier used.

If we consider the Eqs. 1–6, we can find that none of them relates directly the frequency to the results. The channel bandwidth will only increase the number of RBs available to grant to all active users. Other transmission considerations like the use of MIMO will increase the throughput results. The important part is that if we simulate the RR, PF and MT algorithms with the same parameters, MT will give the highest throughput, RR will be the fairest scheduler as it gives resources to all users no matter their channel conditions and PF will try to multiplex as many users as possible while giving a sufficient level of throughput. All these results can be seen in our research and in many other documents like in [12–16], where the numerical outputs are different but the general behavior of the schedulers described before remains the same. Remember that all of these three basic algorithms' behave focus mainly on the throughput result.

6 Conclusions and Future Works

The focus of this paper was to analyze the behavior of Round Robin, Proportional Fair and Maximum Throughput on a 5G NSA network that uses a low frequency band (1800 MHz) when FTP 3 traffic is transmitted. The results show that the Proportional Fair and Maximum Throughput algorithms are most suitable to handle FTP traffic for eMBB services. While the throughput of Round Robin is high for all users (11.79 Mbps compared to the 12.04 Mbps of MT and 3.87 Mbps given by PF in the presence of 50 users), the delay introduced by this scheduler has a negative impact on the PLR (Packet Loss Rate). All RR users dropped between 128 and 245 packets while only some PF users dropped between 0 and 227 packets and some MT users dropped from 0 to 230 packets. These results were obtained with 50 users on the cell (there were any packet dropped from the three algorithms while allocating resources to 5 users).

This research leaves the possibility to design other algorithms based on the three analyzed in this paper. For simplicity, we assume that the whole bandwidth was used only for user plane data transmission. We didn´t handle retransmissions. A simple M/M/1/B process modeled the queue of each user at the gNodeB. Multi-antenna transmissions or schemes like beamforming may enhance the results. A mix of different sources of traffic may require another approach of scheduling, especially if there is real-time traffic or strict QoS requirements. These are only some of the items we can think of right now to improve the scheduling topic for 5G new generation networks. As a final comment, although ITU requirements for eMBB services are not satisfied, we should think that the load, multi-antenna techniques and other factors affect the transmission results.

Acknowledgment. This research was supported by the project No. PIGR-19–06 "Seguridad en comu-nicaciones móviles cooperativas de 5G usando tecnologías de capa física", funded by Escuela Politécnica Nacional.

References

1. Haus, G., Ludovico, A., Pagani, E.: 5G Technology and its applications to music education. Department of Computer Science Giovanni Degli Antoni, pp. 3–4 (2017)
2. 3GPP: Technical Specification Group Radio Access Network; Study on New Radio (NR) access technology (Release 16), pp. 69–70
3. Release 15 - 3GPP: https://www.3gpp.org/release-15. last accessed 1 Sep 2021
4. 3GPP: 3GPP TSG RAN WG1 Meeting #103-e, October 26th – November 13th (2020)
5. Generate and Visualize FTP Application Traffic Pattern: MathWorks. https://la.mathworks.com/help/5g/ug/generate-and-visualize-ftp-application-traffic-pattern.html. last accessed 25 Aug 2021
6. IEEE: 802.11–14/0571r12 - 11ax Evaluation Methodology, pp. 41–42 (2021)
7. Habaebi, M., Chevil, J., Sakkaf, A., Dahawi, T.: Comparison between scheduling techniques in long term evolution (2013)
8. Tetcos: Verification of MAC Scheduling algorithms in NetSim (2014)
9. Sniri, B., Hakimi, W., Mallouki, N.: Performance Comparison of Scheduling Algorithms For Downlink LTE System. University of Tunis El Manar, Tunis-Tunisia (2014)
10. Wang, H., Meng, W., Nguyen, T.: User Fairness Scheme with Proportional Fair Scheduling in Multi-user MIMO Limited Feedback System. Harbin Institute of Technology, Harbin, China (2013)
11. ITU: Preliminary Evaluation Report from the 5G infrastructure association on IMT-2020 proposal (2019)
12. Elshennawy, M.: Modified Proportional Fair Scheduling Algorithm for Heterogeneous LTE-A Networks (2020)
13. Armour, SMD., Beh, K.C.: Joint time-frequency domain proportional fair scheduler with HARQ for 3GPP LTE systems (2008)
14. Liu, E., Leung, K.: Fair Resource Allocation under Rayleigh and/or Rician Fading Environments. Department of Electrical and Electronic Engineering Imperial College (2008)
15. Anand, A., de Veciana, G., Shakkottai, S.: Joint Scheduling of URLLC and eMBB Traffic in 5G Wireless Networks. Department of Electrical and Computer Engineering, The University of Texas at Austin (2018).august
16. Perdana, D., Nur Sanyoto, A., Gustommy, Y.: Performance Evaluation and Comparison of Scheduling Algorithms on 5G Networks using Network Simulator (2019)
17. Zhu, R., Yang, J.: Buffer-aware adaptive resource allocation scheme in LTE transmission systems. EURASIP J. Wirel. Commun. Netw. **2015**(1), 1–16 (2015). https://doi.org/10.1186/s13638-015-0398-y
18. Walpole, E.: Probability and Statistics for Engineers and Scientists, 8th edition. Pearson (2007)
19. Monroe, W.: The Normal Distribution, pp. 1–2 (2017)
20. Gebali, F.: Analysis of Computer and Communication Networks. Springer (2008)
21. 3GPP: RP-090746_Annex C2 Link budget template NLOS (2009)
22. Jing-Hee, C., Jing-Ghoo, C., Chcuk, Y.: Analysis of Packet Transmission Delay Under the Proportional Fair Scheduling Policy (2006)
23. 3GPP: 5G; Study on scenarios and requirements for next generation access technologies (2017)

A Comparative Analysis of External Optical Modulators Operating in O and C Bands

Abigail Rivadeneira(✉), María Soledad Jiménez, and Felipe Grijalva

Escuela Politécnica Nacional. Ladrón de Guevara, 11-235 Quito, Ecuador
{susana.rivadeneira,maria.jimenez,felipe.grijalva}@epn.edu.ec

Abstract. These days, optical fiber is the preferred transmission medium and the global backbone of the internet. New technologies for optical communication systems have been developed to fulfill the rapidly increasing demand, including external optical modulation. This type of modulation enables high speeds and long-distance operation. Some external modulators widely used in high-performance optical communication systems are electro-optical modulators and electro-absorption modulators. The purpose of this study is to compare the performance of external electro-absorption and electro-optical modulators under different scenarios of single-channel optical systems operating in O and C bands. Towards that end, it is used OptiSystem simulation software. The simulation scenarios were carried out by changing parameters that alter the communication system's performance, such as link length and transmission speed. The analysis shows that electro-optical modulators are superior to electro-absorption modulators in terms of Bit Error Rate for both O and C bands.

Keywords: External optical modulation · Electro-absorption modulator · Electro-optical modulator

1 Introduction

Fiber-optic communication systems have made significant advances since the 1970s [1]. Currently, optical fiber has positioned itself as the preferred transmission medium in long-haul communications systems, becoming the global backbone of the Internet [2].

The most outstanding advantages of fiber-optic communication systems over copper are wide bandwidth, signal transmission over long distances, security, immunity to electromagnetic interference, and low signal attenuation [3]. The advantages offered by optical fiber make it an ideal transmission medium for new and increasingly demanding applications.

New technologies have been developed for optical communication systems to meet the growing demand for high speeds. External optical modulation is part of these technologies and is key to future communications over optical fiber.

Nowadays, external modulators are widely used in high-performance systems. Over the last decade, waveguide based LiNbO3 modulators have been the basis for generating

M. Botto-Tobar et al. (Eds.): *Latest Advances in Electrical Engineering, and Electronics*, LNEE 933, pp. 219–232, 2022.
https://doi.org/10.1007/978-3-031-08942-8_16

advanced modulation formats. Although conventional external optical modulators are reaching their performance limit because they are bulky and consume lots of power, new types of external optical modulators are emerging. These improved external modulators change the materials they are made of (such as LiNbO3 over insulator known as LNOI), but their operating principle remains the same [4].

Recent studies attempt to improve the performance of electro-absorption and electro-optical modulators by varying their materials. Concerning electro-optical modulators, a limitation was the inability to be monolithically integrated into the optical source since they were made of crystals. Some works analyzed the possibility of using semiconductors [5] to reduce power consumption and cost, and others the use of graphene [6] to improve the performance of these modulators. Later works analyzed the possibility of re-using crystals, specifically lithium niobate, for being one of the best electro-optical materials [7]. Recent work [8, 9] assessed the possibility of combining the advantages of both semiconductors and crystals by using silicon photonic integrated circuits with lithium niobate thin films.

Similarly, the study over recent years of electro-absorption modulators focuses on varying the manufacture materials to improve the characteristics of these modulators. Some works employ heterogeneous germanium and silicon structures to improve power consumption in new generations [10, 11]. Recent work analyses the use of graphene in electro-absorption modulators to improve their performance and modulation efficiency [12, 13].

Several studies have compared the performance of electro-optic modulators with electro-absorption modulators. For example, the work of Pelusi compares the performance of these modulators in OTDM (Optical Time Division Multiplexing) systems [14], Latif et al. compare them in systems using ROF (Radio Over Fiber) [15], and Hazem et al. compare the performance of these modulators in OW (Optical Wireless) channels [16]. What is noteworthy is that they all reach the same conclusion, which coincides with that of this work, electro-optical modulators perform better than electro-absorption modulators.

In contrast to previous works, this work emphasizes the BER performance in a single channel wired optical communication system for a baseband signal for each type of external modulator (i.e., electro-optical and electro-absorption) rather than their compositions. We also compare the modulators' configurations of electro-optic and electro-absorption modulators empirically through simulations. Specifically, this study seeks to analyze the behavior of high-performance external optical modulators by simulating single-channel optical systems in O and C bands.

2 External Optical Modulators

Modulators are part of the optical transmitter. Their function is to vary at least one parameter of an optical carrier wave according to the electrical signal containing information. If the modulator regulates the current injected into the light source, it is called direct modulation. If the modulator modifies the characteristics of constant light flux, it is called external modulation.

External modulators work at higher speeds and longer distances. The separation of the optical source and modulator helps to reduce the frequency chirp. Despite the advantages of external optical modulation, it is more expensive than direct modulation and requires more complex circuitry to handle the high-frequency signal.

There are three types of external optical modulators: acousto-optic modulators, electro-optic modulators, and electro-absorption modulators. This paper focuses on the electro-optical and electro-absorption modulators, which are currently the most widely used in high-performance optical communication systems. Next, we describe both of them.

2.1 Electro-Optical Modulator

The electro-optical modulator applies an electrical signal that causes a change in the refractive index of the modulator's material. This change determines the amount of power that couples from the input to the output. The device itself consists of either an optical waveguide or an MZI [1].

Depending on its configuration, the electro-optical modulator modulates in phase, intensity, or complex optical field. The phase electro-optical modulator is the simplest and it is used as a building block for other electro-optical modulators. It consists of an optical waveguide and a pair of electrodes responsible for inducing an electric field.

For intensity modulation, it uses an MZI. In this configuration, the optical signal splits into two beams crossing the interferometer arms. Depending on the disposition, one or both paths of the MZI have a pair of electrodes. The electric field between the electrodes controls the phase between the split signals. Thus, the split signals interfere constructively or destructively at the output. The electro-optical intensity modulator has DP-M, DD-MZ-AP, and SD-MZ configurations.

The I/Q modulator improves the transmission capacity by carrying information in both phase and amplitude. This modulator uses three MZI, two independent and one dependent. The independent ones are placed one in each of the arms of the third dependent MZI. In addition, are added a pair of electrodes to one of the arms of the dependent MZI to introduce a phase modulation.

2.2 Electro-Absorption Modulator

The electro-absorption modulator applies an electric field to a semiconductor. When light passes through it, a part of it is absorbed [17]. This modulator, manufactured with semiconductor materials, can be integrated with lasers into the same package making it an attractive alternative [1].

The external electro-absorption modulator just modulates in intensity since part of the energy is absorbed. However, the application of an electric field induces the electro-optic effect too and produces a phase modulation. If the electro-optic effect is strong enough to cause excessive phase modulation, it results in a frequency chirp [18].

3 Simulation Scheme

To compare the performance of the modulators were used simulations of optical communication systems. The simulation tool used for this purpose is OptiSystem version 17 from OptiWave.

The simulations with both modulators take place in single-channel systems to focus on the behavior of these modulators. These simulations compare the performance of the systems with the modulators by varying parameters such as link length, transmission power, and transmission speed.

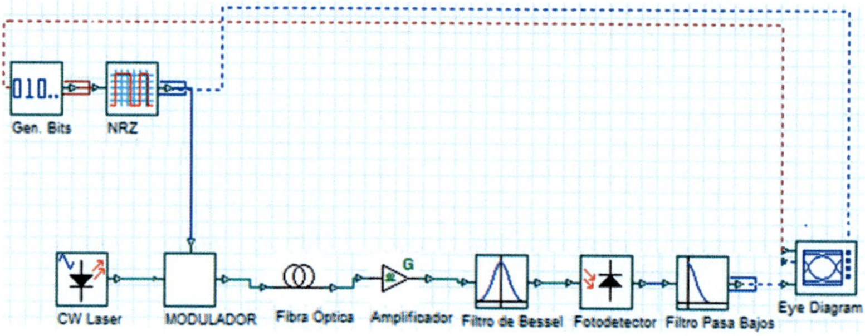

Fig. 1. General scheme of the simulation scenarios.

Figure 1 shows the general scheme of the simulations. Next, we give a brief description of the components used.

3.1 Transmitter

Bit generator. A bit generator is used at transmission rates from 100 Gbps to 250 Gbps which are commonly used speeds in single-channel optical systems.

Encoder. NRZ (Non-Return to Zero) coding is used in the bitstream before modulation.

Optical Source. A generic source of OptiSystem's software is used to generate a continuous light flux in O and C bands, two bands commonly used in optical systems. We use two output power values 5 dBm and 7 dBm. Considering that the output power values commonly used in practice are between 4 and 6 dBm, and since the simulator modulators include losses, we chose 5 dBm and 7 dBm to give the simulator systems a practical approach. The 7 dBm transmission power is only used in systems that require it, such as 100 km links or longer-range links that are divided into 100 km sections. The 5 dBm power is used in links smaller than 100 km or links with sections smaller than 100 km.

3.2 Modulators

The OptiSystem transmitter library offers both electro-absorption and electro-optical modulators. The library consists of two variants of the electro-absorption modulator and five of the electro-optical modulator. These modulators are presented in Table 1.

The electro-absorption modulators available in the software differ in that the EA-A allows the parameter values to be set flexibly without being affected by the voltages

Table 1. External modulators.

Modulator	Labeled as
EA Modulator Analytical	EA-A
EA Modulator Measured	EA-M
MZ Modulator Analytical	MZ-A
Dual Drive MZ Absorption-Phase	DD-MZ-AP
Single Drive MZ Modulator	SD-MZ
Dual Port Dual Drive MZ Modulator Absorption Phase	DPDD-MZ-AP
Dual Port Modulator Measured	DP-M

applied in the modulator. On the other hand, the EA-M has a more realistic behavior since it depends on the applied reverse bias voltage.

On the other hand, the electro-optical modulators available in the software are only amplitude modulators and differ in the configuration they use. The MZ-A is the simplest one since it does not consider losses due to absorption of the optical signal or insertion to the optical source. In addition, the MZ-A has an inherent chirp.

The SD-MZ modulator has the peculiarity that one of the voltages in one arm of the MZI is zero. This modulator considers the dependence between absorption losses and the phase of the applied voltage.

The DD-MZ-AP considers the dependence between the absorption losses and the phase change produced by the applied voltage. This consideration results in nonlinearities in the modulation if the chosen working point is not correct. In addition, this modulator can control the frequency chirp.

The DPDD-MZ-AP considers two different voltage inputs for each arm of the MZI.

The Dual Port MZ Modulator Measured electro-optical modulator considers two different voltage inputs for each arm of the MZI. What differentiates it from the DPDD-MZ-AP is that it does not consider the relationship between absorption losses, phase, and applied voltage. However, it contemplates parameters that no other electro-optic modulator allows specifying, like switching voltages and insertion losses to the optical source.

3.3 Transmission Channel

The simulations are performed at O and C bands since these are the bands in which these modulators commonly work. The simulated links exceeded 100 km to observe the performance of the modulators in demanding systems to work at their limits. This value falls within the values used in practice since a long-haul fiber-optic link can range from a few hundred to several thousand kilometers [19].

O-band. In the case of the 1310 nm O-band, the parameters of the standard single-mode SMF-28® ULL Optical Fiber from Corning [20] are used as a reference since typically optical systems use this fiber to work at this wavelength.

C-band. For 1550 nm C-band are used parameters of the single-mode LEAF® Optical Fiber from Corning [21]. This fiber is typically used at this wavelength because it has a large effective area that reduces nonlinear effects. LEAF fiber is an NZDSF (Non-Zero Dispersion-Shifted Fiber), so to compensate for the effects of chromatic dispersion, it is alternated between NZD + and NZD- fibers.

Line Amplifiers. Line amplifiers ensure that the optical signal level remains above the noise floor. They are used on links exceeding 100 km in length. In the simulations, the link is divided whether into 100 km or 80 km sections. According to this, 20 dB and 16 dB line amplifiers are used, respectively, with noise figures of 4 dB.

3.4 Receiver

Preamplifiers. In the receiver section, because the signal arrives weakened, preamplifiers are used to amplify the signal. The simulations use a preamplifier with 20 dB of gain and a noise figure of 4 dB.

Photodetector. The filtered signal passes through a PIN photodetector which converts the optical signal to electrical.

Filters. In the simulations, two filters are applied. A pass-band filter to reduce ASE (Amplified Spontaneous Emission) of the optical signal and a low-pass filter is applied after optical detection to remove unwanted components of the electrical signal. Finally, we evaluate the system's performance with an eye diagram tool to calculate the system's BER.

4 Results Analysis

4.1 Results with the Electro-Optical Modulators

In Fig. 2, it is possible to observe the performance of each electro-optical modulator available in the software for a link of length L = 100 km operating at 1550 nm and a fixed transmit power (Ptx) of 7 dBm by plotting the BER of the system as a function of the transmission rate. The data for this plot was obtained by simulations with different transmission rates in OptiSystem software.

From Fig. 2, DD- P is the modulator that obtains the best BERs for most of the speeds used. In the C-band, this modulator has the best performance since it controls the frequency chirp.

As second of best performance in the figure is the MZ-A. This modulator has good performance because although it has an inherent chirp in its configuration. However, it is worth noting that unlike other electro-optical modulators available in the software, the MZ-A does not consider absorption losses, phase shifts due to the applied voltage, or insertion losses to the optical source, making this modulator quite ideal. In addition, the ER (Extinction Ratio) of this modulator can be set, and for the simulations, it was set to 20 dB.

The third and fourth modulators with the best performance are the SD-MZ and DPDD-MZ-AP, respectively. These two modulators consider the same absorption losses and phase shifts by the applied voltage. In addition, these two configurations do not control the frequency chirp, making them quite similar. The main difference between these modulators that gives an advantage to the SD-MZ modulator over the DPDD-MZ-AP modulator is the ER. The SD-MZ modulator achieves an ER of 19.54 dB, while the DPDD-MZ-AP modulator achieves an ER of 17.31 dB.

Finally, the modulator with the worst performance is the DD-MZ-AP configuration. This modulator, like the DPDD-MZ-AP, does not control the chirp. However, it allows having a theoretical ER of infinite value.

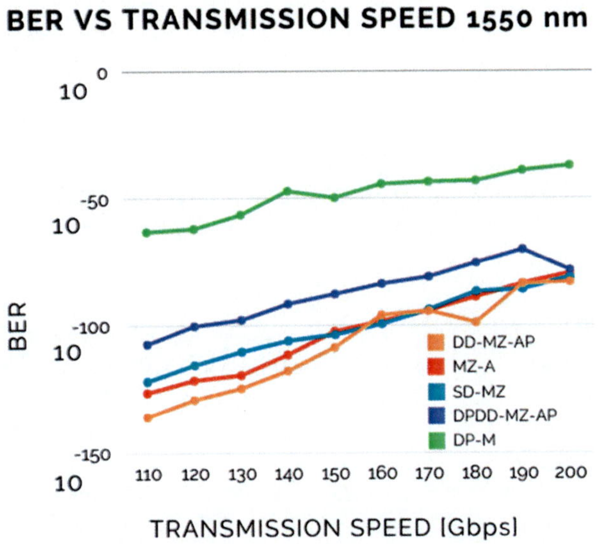

Fig. 2. BER vs Transmission Speed for an optical communications system with $\lambda = 1550$ nm, Ptx = 7 dBm and, L = 100 km for different electro-optical modulators.

The major limitation of this modulator is that it considers the insertion loss between the modulator and the optical source, which in the case of the simulations performed is 5 dB. One limitation of electro-optical modulators is the insertion loss because depending on the material of manufacture they can or not be integrated into optical sources.

Other technologies of electro-optic modulators created from semiconductor materials allow monolithic integration [22]. However, conventional modulators manufactured from crystals do not. In this case, the Dual Port modulator considers insertion losses to the source. If a semiconductor modulator was considered, losses should be zero (as for the other electro-optic modulators available in the software).

The results obtained are reinforced with a BER vs. Distance graph presented in Fig. 3. With this graph, it is possible to corroborate that the performance order of the modulators follows the previously obtained arrangement, being the DD-MZ-AP the one with the best BER for a given distance.

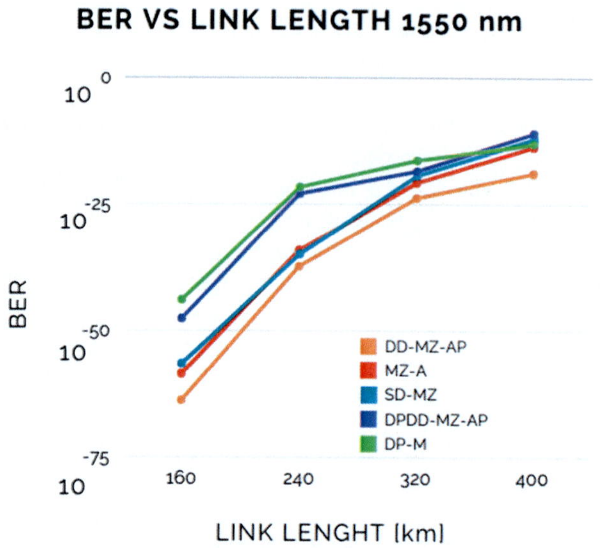

Fig. 3. BER vs Link Lenght for an optical communications system with 80 km link sections, Ptx = 5 dBm, 16 dB line amplifiers, λ = 1550 nm and, Vt = 200 Gbps for different electro-optical modulators.

In Fig. 4, it is possible to observe the performance of each electro-optical modulator available in the software for a 100 km link operating at 1310 nm for transmission power of 7 dBm by plotting the BER of the system as a function of the transmission speed (VT). The data for this plot was obtained by simulations with different transmission rates in OptiSystem software.

It is evident that the performance of the modulators at the O-band changes concerning the C-band for the same modulation scenarios. The MZ-A, SD-MZ, and DP-M improve their performance concerning that achieved at the C-band. This change is basically because these three modulators were minimally affected by the phase differences that define the constructive or destructive interference at the output of the MZI.

The MZ-A achieved 98.2% of its maximum output amplitude for logic 1 and exceeded the amplitude for logic 0 by 0.98%. The SD-MZ modulator achieved 82.8% of its maximum output amplitude for logic 1 and exceeded the amplitude for logic 0 by 0.92%. The Dual Port modulator reached 100% of its maximum amplitude for logic 1 and did not exceed the amplitude for logic 0.

These changes benefit the three mentioned modulators to improve their performance at the O-band rather than C-band. By having a clear difference between the amplitude of logic 0 and logic 1, the noise attenuation at the receiver improves the system's BER.

Conversely, the DD-MZ-AP and DPDD-MZ-AP modulators do not benefit since the difference between the amplitude of logic 0 and logic 1 is not strong. The DD-MZ-AP modulator reached 69.6% of its maximum output amplitude for logic 1 and did not exceed the amplitude for logic 0. The DPDD-MZ-AP modulator reached 77.6% of its maximum output amplitude for logic 1 and exceeded the amplitude for logic 0 by 1.44%.

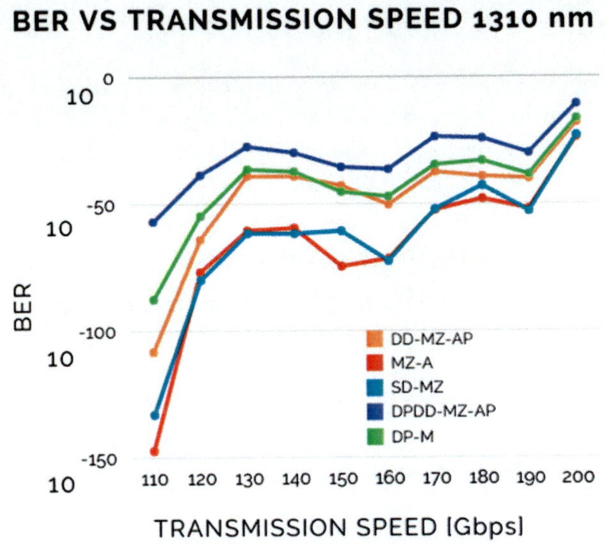

Fig. 4. BER vs Transmission Speed for an optical communications system with $\lambda = 1310$ nm, Ptx = 7 dBm and, L = 100 km for different electro-optical modulators.

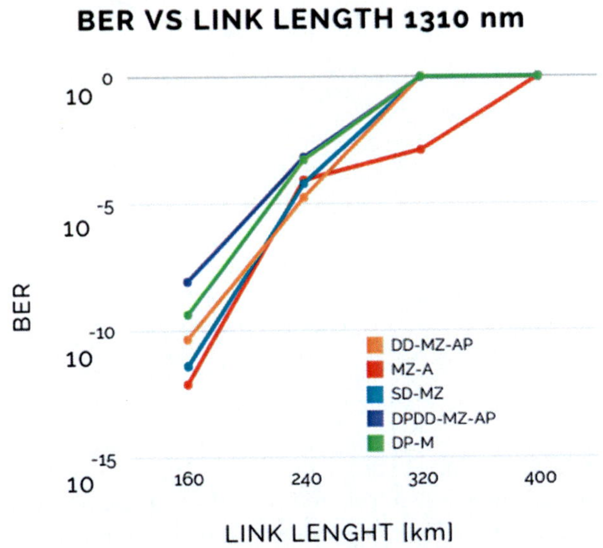

Fig. 5. BER vs Link Lenght for an optical communications system with 80 km link sections, Ptx = 5 dBm, 16 dB line amplifiers, $\lambda = 1310$ nm and, Vt = 200 Gbps for different electro-optical modulators.

This blurred difference makes the noise attenuation harm the system's BER with these modulators.

The results obtained in the O-band are reinforced with a BER vs. distance graph presented in Fig. 5. In this figure, it is possible to corroborate that the performance order of the modulators follows the previously obtained, being the MZ-A modulator the one with the best BER for an established distance.

4.2 Results with the Electro-Absorption Modulators

Figure 6 shows the performance of each electro-absorption modulator available in the software for a 100 km link operating at 1550 nm for a fixed transmit power of 7 dBm by plotting the BER of the system as a function of the transmission rate. The data for this plot was obtained by simulations with different speeds in OptiSystem software.

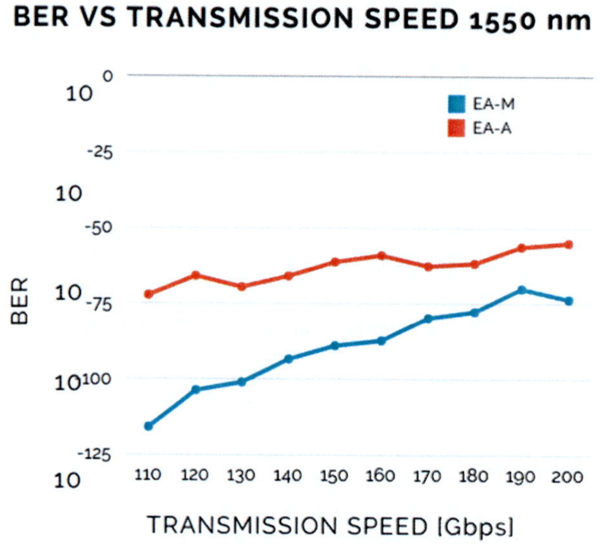

Fig. 6. BER vs Transmission Speed for an optical communications system with $\lambda = 1550$ nm, Ptx = 7 dBm, and L = 100 km for different electro-absorption modulators.

The EA-M has a superior performance mainly due to the improvement in the modulation index in this modulator. In addition, another parameter that collaborates in the performance improvement of this modulator is the significant decrease of the chirp in frequency due to the bias voltage used.

It is worth noting that the EA-M achieves a good performance over the EA-A even though it considers the dependence between the applied voltage, the absorption of the light signal. If the EA-A's parameters were ideal, such as a small frequency chirp and a modulation index of one, the scenario would change, and the EA-A would have a better performance.

EA-A can radically change its performance depending on the value of the parameters that are chosen. This modulator allows a flexible set of values, unlike the EA-M.

The EA-M is the most realistic approximation of an electro-absorption modulator that OptiSystem offers. This modulator has a performance like the Dual Port MZ Modulator, which is the only electro-optical modulator available in OptiSystem that considers insertion losses between the modulator and the optical source.

This shows that an important limitation of crystal-based electro-optical modulators is the insertion loss to the optical source because they cannot be monolithically integrated into the optical source. This problem makes their performance decrease and is comparable to electro-absorption modulators. As mentioned previously, this problem can be solved by employing semiconductor-based electro-optical modulators.

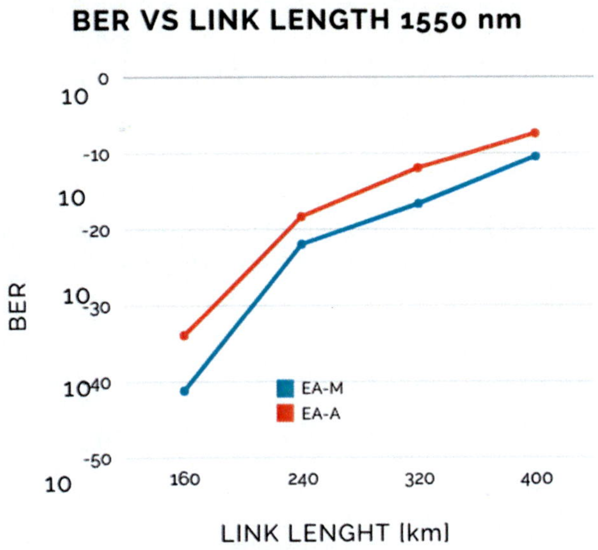

Fig. 7. BER vs Link Lenght for an optical communications system with 80 km link sections, Ptx = 5 dBm, 16 dB line amplifiers, $\lambda = 1550$ nm, and Vt = 200 Gbps for different electro-absorption modulators.

The results obtained in the C-band are reinforced with a BER vs. link length presented in Fig. 7. Observe that the performance order of the modulators follows the previously obtained arrangement, being the EA-M the one with the best BER for an established distance.

In Fig. 8, it is possible to observe the performance of each electro-absorption modulator available in the software for a 100 km link operating at O-band for transmission power of 7 dBm by plotting the BER of the system as a function of the transmission speed (VT). The data for this plot was obtained by simulations with different transmission rates in OptiSystem software.

As in the C-band, the performance of the EA-M is better than that of the EA-A since it achieves a better BER for a given transmission rate. This happens because of the significant decrease in the frequency chirp and the improvement in the modulation index in the EA-M.

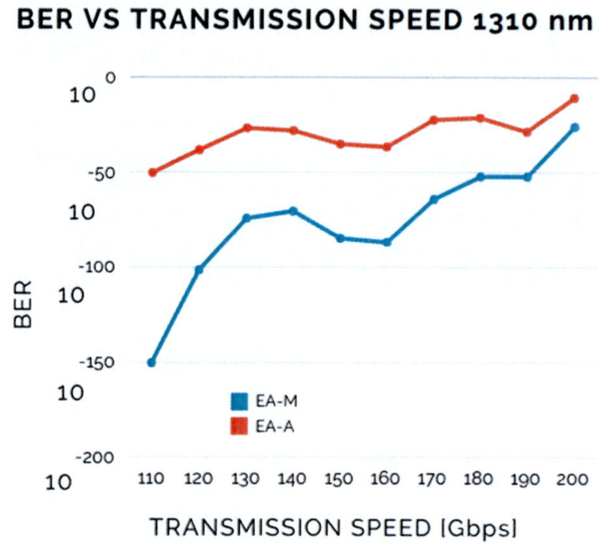

Fig. 8. BER vs Transmission Speed for an optical communications system with $\lambda = 1310$ nm, Ptx = 7 dBm, and L = 100 km for different electro-absorption modulators.

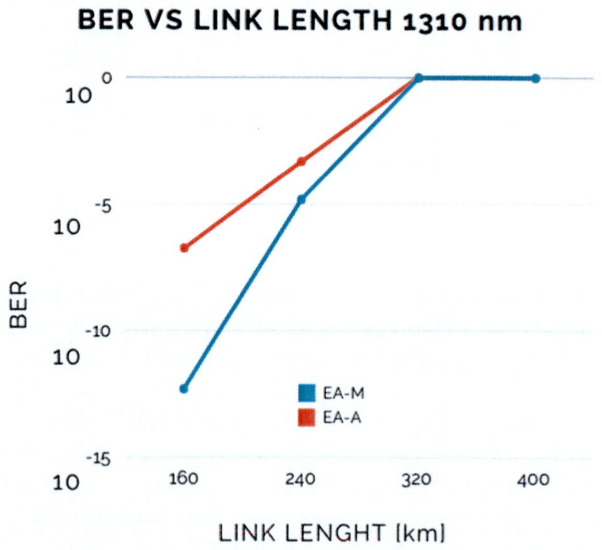

Fig. 9. BER vs Link Lenght for an optical communications system with 80 km link sections, Ptx = 5 dBm, 16 dB line amplifiers, $\lambda = 1310$ nm, and Vt = 200 Gbps for different electro-absorption modulators.

The results obtained in the O-band are reinforced with a BER vs. distance presented in Fig. 9. Note that the performance order of the modulators follows the previously obtained arrangement, being the EA-M the one with the best BER for a given distance.

5 Conclusions

Through the simulations of the high-performance systems with both modulators, it was possible to observe the performance of the studied modulators in terms of Bit Error Rate. The electro-optical modulators maintained a better performance than the electro-absorption modulators in the O and C bands.

It was possible to verify by simulation that frequency chirp is a parameter that substantially limits the performance of optical communication systems. Electro-absorption modulators, which always have a frequency chirp, maintain high Bit Error Rates in the C band. This occurs because the electro-absorption effect is always accompanied by electro-optical effects so that variations in intensity produce variations in the signal phase that in turn cause the frequency chirp. Electro-optical modulators, on the other hand, can control chirp and even eliminate it thus, they maintain better Bit Error Rates in this band.

In the O-band, one of the electro-absorption modulators (EA-M) improves its Bit Error Rate significantly compared to the C band, becoming the modulator with the best performance. This happens because this modulator has a good extinction ratio, which benefits it in the O-band due to the attenuation that characterizes that band.

For future work, it is feasible to study the mentioned configurations using the characteristics of the new emerging materials such as graphene, silicon photonic integrated circuits with lithium niobate thin-films, and heterogeneous structures of germanium and silicon. Combining both aspects will provide better characteristics for future networks.

Similarly, an extension of this work could analyze from the point of view of other parameters such as OSNR or Q-Factor.

References

1. Ramaswami, R., Sivarajan, K., Sasaki, G.: Optical Networks: A Practical Perspective. Morgan Kaufmann (2009)
2. Hui, R.: Introduction to Fiber-Optic Communications. Academic Press (2019)
3. Massa, N.: Fiber optic telecommunication. Fundam. photonics 298 (2000)
4. Xu, M., et al.: High-performance coherent optical modulators based on thin-film lithium niobate platform. Nat. Commun. **11**, 1–7 (2020)
5. Abdelatty, M.Y., Badr, M.M., Swillam, M.A.: Compact silicon electro-optical modulator using hybrid ITO tri-coupled waveguides. J. Light. Technol. **36**, 4198–4204 (2018)
6. Zhou, M., et al.: Ultrawide bandwidth and sensitive electro-optic modulator based on a graphene nanoelectromechanical system with superlubricity. Carbon N. Y. **176**, 228–234 (2021)
7. Li, M., Ling, J., He, Y., Javid, U.A., Xue, S., Lin, Q.: Lithium niobate photonic-crystal electro-optic modulator. Nat. Commun. **11**, 1–8 (2020)
8. Boynton, N., et al.: A heterogeneously integrated silicon photonic/lithium niobate travelling wave electro-optic modulator. Opt. Express **28**, 1868–1884 (2020)

9. Ahmed, A.N.R., Nelan, S., Shi, S., Yao, P., Mercante, A., Prather, D.W.: Subvolt electro-optical modulator on thin-film lithium niobate and silicon nitride hybrid platform. Opt. Lett. **45**, 1112–1115 (2020)
10. Mastronardi, L., et al.: High-speed Si/GeSi hetero-structure electro absorption modulator. Opt. Express **26**, 6663–6673 (2018)
11. Feng, N.-N., et al.: 30GHz Ge electro-absorption modulator integrated with 3μm silicon-on-insulator waveguide. Opt. Express **19**, 7062–7067 (2011)
12. Fu, G., et al.: A compact electro-absorption modulator based on graphene photonic crystal fiber. Chinese Phys. B **29**, 034209 (2020)
13. Chakraborty, I., Roy, S., Dixit, V., Debnath, K.: Atto-joule energy-efficient graphene modulator using asymmetric plasmonic slot waveguide. Photonics Nanostructures-Fundamentals Appl. **43**, 100865 (2021)
14. Pelusi, M.D.: Fiber-looped LiNbO3 mach-zehnder modulator for 160 Gb/s optical time division demultiplexing and it's comparison to an electro-absorption modulator. OFC/NFOEC 2008 - 2008 Conf. Opt. Fiber Commun. Fiber Opt. Eng. Conf. **1**, 79–81 (2008)
15. Latif, A., Hussain, A., Khan, F., Hussain, A., Khan, Y., Munir, A.: A performance based comparative analysis of high speed Electro Absorption and Mach-Zehnder Modulators to mitigate chromatic dispersion at 140 GHz millimeter wave. Adv. Inf. Sci. Serv. Sci. **4**(20), 368–377 (2012)
16. El-Hageen, H.M., Kuppusamy, P.G., Alatwi, A.M., Sivaram, M., Yasar, Z.A., Zaki Rashed, A.N.: Different modulation schemes for direct and external modulators based on various laser sources. J. Opt. Commun. no. Dc, 1–10 (2020)
17. Agrawal, G.P.: Fiber-optic communication systems. John Wiley & Sons (2012)
18. Hunsperger, R.G.: Integrated Optics: Theory and Technology. Springer (2009)
19. Corning: Long-haul Terrestrial Optical Fiber Networks. [Online]. Available: https://www.corning.com/optical-communications/worldwide/en/home/applications/long-haul-networks.html
20. Corning: Corning® SMF-28® Ultra optical fiber. [Online]. Available: https://www.corning.com/media/worldwide/coc/documents/Fiber/SMF-28Ultra.pdf
21. Corning: Corning® LEAF® Optical Fiber. [Online]. Available: https://www.corning.com/media/worldwide/coc/documents/Fiber/LEAFopticalfiber.pdf
22. Dagli, N.: High-Speed Photonic Devices. CRC Press, Boca Raton (2006)

Logarithmic Antennas for Electromagnetic Energy Harvesting Applications

Carlos Gordón(✉), Evelyn Freire, Geovanni Brito, and Fabian Salazar

GITED Research Group, Facultad de Ingeniería en Sistemas, Electrónica e Industrial, Universidad Técnica de Ambato, UTA, 180150 Ambato, Ecuador
{cd.gordon,el.freire,geovannidbrito,fr.salazar}@uta.edu.ec

Abstract. The main objective was the realization and implementation of two designs of spiral and periodic logarithmic antennas, in order to test the feasibility of this technology by storing energy in rechargeable batteries in different environments with the presence of WiFi networks using coaxial cable conductors (copper), optical fiber and an environment with offset antennas. Different existing antennas were investigated and modifications were made to work in the 2.4 GHz band in order to adjust them to our objectives. For this purpose, CST Studio 2018 simulation software was used to perform the designs and review the parameters emitted by each antenna. Finally, the antennas were built based on the proposed design, as well as a voltage multiplier circuit with Schottky diodes and finally, performance tests and validation of the prototype were carried out using the miniVNA Tiny spectrum tester.

Keywords: Logarithmic antennas · Energy harvesting · Spiral antenna · Periodic antenna

1 Introduction

Alternative energy is defined as an energy source that "is an alternative to the use of fossil fuels with a low environmental impact" [1]. In recent years, technology has evolved significantly in the invention and development of alternatives to reduce the consumption of fossil fuels, generating clean and renewable energy, where research and applications of devices have been carried out to perform a process of energy transformation, recovering it, storing it and then reusing it. This system has been called Energy Harvesting [2].

With the increase of the population and the development of the economy, the topic of clean energies is transcendental and increasingly used in the research area, since it is vital to make up for the shortage of energy resources. For this reason, different energy sources such as solar, wind, nuclear, biomass, hydro and also RF energy harvesting are being sought [3]. RF energy has been widely investigated during the last 50 years, and there has been an increase in communications systems that use the electromagnetic spectrum through the design of systems that allow capturing a greater amount of energy by means of patch antennas with rectifier [4].

M. Botto-Tobar et al. (Eds.): *Latest Advances in Electrical Engineering, and Electronics*, LNEE 933, pp. 233–249, 2022.
https://doi.org/10.1007/978-3-031-08942-8_17

Currently, the use of ambient energy to generate electrical energy has boosted the applications in which Energy Harvesting provides service because it is an ecological and autonomous energy that can be used in countless applications, such as powering small sensors that do not require large amounts of energy for their operation [5]. This form of energy, in combination with the new technologies that have emerged in recent years, makes us rethink the new ways of powering wireless devices [5]. There are some projects and works related to the design of antennas for different wireless power transmission applications, mostly consist of a rectifier antenna which receives a radio frequency wave and converts it into DC power that is used in wireless power systems for applications based on system autonomy [5]. A comprehensive study of different radio frequency energy harvesting (RFEH) systems such as rectifier circuit design and matching networks is presented, and finally a general framework for ambient RFEH system design is deduced [6]. There are suitable antennas for energy harvesting, 4 basic structures are discussed which are patch antenna, slot antenna, modified inverted F antenna and dielectric resonator antenna, operating between 0.8 GHz and 2.6 GHz, i.e., GSM, UMTS and WiFi [8, 9].

In this paper, we present the implementation of 2 antennas. Logarithmic Spiral Antenna (LSA) and Logarithmic Periodic Antenna (LPA) for electromagnetic energy harvesting applications. As for the manufacture of the prototype, it was carried out at Electrodrone [7]. Both antennas were designed at a frequency of 2.4 GHz. (802.11 n) because this band is the most common in our environment. This article is constituted as follows: the design, materials and methodology used in the development of the proposed antennas are described in Sect. 2. The simulation and measurement results along with the discussion are provided in Sect. 3.

2 Methodology

This section presents the methodology used for the development of logarithmic antennas for electromagnetic energy collection. The proposed antennas were designed with CST Studio software. The design of 2 antennas is proposed: spiral logarithmic and periodic logarithmic at a central frequency of 2.4 GHz.

2.1 Logarithmic Spiral Antenna (LSA) Design

- **Initial frequency parameters**

Table 1. LSA initial frequency

Parameter	Value	Unit
Center frequency	2.4	GHz
Wavelength (γ)	0.125	Mm

- **Design parameters**

Table 2. LSA design parameters

Parameter	Abbreviation	Re-escalation	Redesigned	Unit
Inside radius	Rin	14.76	0.33	mm
External radius	Rout	16.56	4.8	mm
Number of laps	N	0.25	0.25	-
Characteristic Impedance	Z0(conector)	50	50	Ohms

- **Design parameters for the impedance adapter**

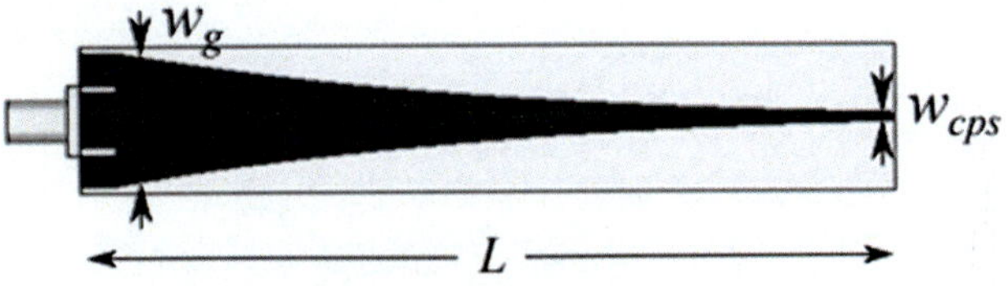

Fig. 1. Impedance adapter model.

Table 3. Parameters for impedance adapter

Parameter	Abbreviation	Value	Unit
Load	L	125	mm
Microstrip width	Wi	5	mm
Ground plane width GND	Wg	18	mm

Table 2, shows the values of the impedances that will be used for the simulation and to obtain the sizing of the impedance adapter (Fig. 1, Table 3 and Table 4).

- **Calculation of effective length from the equation** [11]

$$\varepsilon_{\mathrm{eff}} = \frac{\varepsilon_{\mathrm{r}} + 1}{2} \tag{1}$$

where:

$\varepsilon_{\mathrm{eff}}$ = effective dielectric constant
ε_{r} = relative permittivity

$$\varepsilon_{\mathrm{eff}} = \frac{4.23 + 1}{2}$$

$$\boldsymbol{\varepsilon_{eff} = 2.615}$$

$$c = \frac{c}{(\varepsilon_r + 1)^{0.5}} \tag{2}$$

where:
c = speed light

$$\mathbf{c = 185517724,2}$$

$$R_{out} = \frac{c}{2\pi f_{low}} \tag{3}$$

where:
R_{out} = outside radius.
f_{low} = low frequency

$$\mathbf{R_{out} = 14,76\,mm}$$

$$L = \frac{c}{2\pi f_{low}} \tag{4}$$

where:
L = arm length
f_{low} = low frequency

$$\boldsymbol{L = 14,76\,mm}$$

$$R_{in} = \frac{c}{2\pi f_{high}}$$

where:
R_{in} = inside radius
f_{low} = high frequency

$$R_{in} = \frac{185517724,2}{2\pi \times 2.8\,GHz}$$

$$\mathbf{R_{in} = 16,56\,mm}$$

$$Z_{in} = \frac{120\pi}{2\left(\varepsilon_{eff}^{0.5}\right)} \tag{5}$$

where:
Z_{in} = input impedance

$$\mathbf{Z_{in} = 116,564\,Ohms}$$

$$\mathbf{a} = 0.1$$

- **Simulation of the Logarithmic Spiral Antenna**

The prototype designed for analysis and simulation was made with CST Studio software, the geometry is shown at Fig. 2.

The antenna was designed for a center frequency of 2.4 GHz.

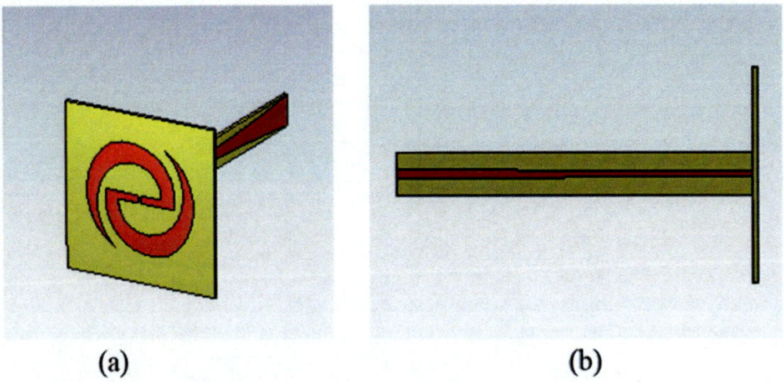

Fig. 2. Simulation for LSA (a) front view, (b) cross-sectional view.

S11 parameters depicted in Fig. 3 or also known as scattering parameters are analyzed to characterize a linear device of one or more ports, in this case a discrete port.

The S11 parameter measures the amount of reflected power and the power applied to the port.

The S11 parameter of the spiral logarithmic antenna is shown in the figure, where a red line points to negative infinity indicating that the antenna is optimal for working in the 2.4 GHz band, thus this model is suitable for implementation because it works at the desired frequency.

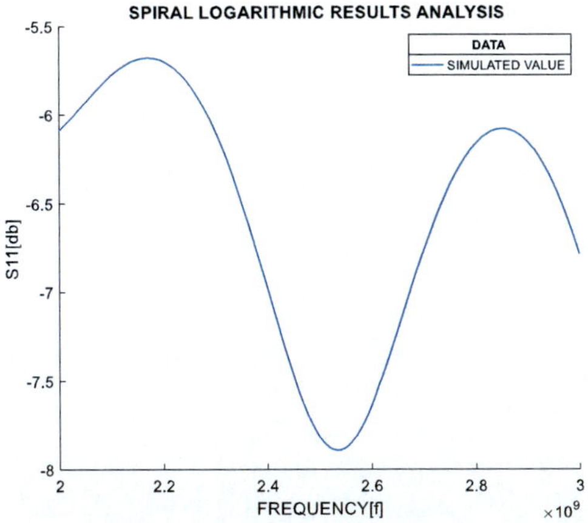

Fig. 3. Parameters S11 of LSA.

- **Manufacturing of the Logarithmic Spiral Antenna**

The spiral logarithmic antenna design, whose ground structure and patches are printed on each side of the FR4 dielectric substrate. The front and adapter of the fabricated prototype of the microband line-fed antennas are illustrated in Fig. 4.

The side shows the metal part printed on the entire antenna geometry, soldered to a 50 Ohms SMA connector, which is used to connect with RG58 coaxial cable to the MiniVNA-TINY, which will allow us to obtain real data from the antenna.

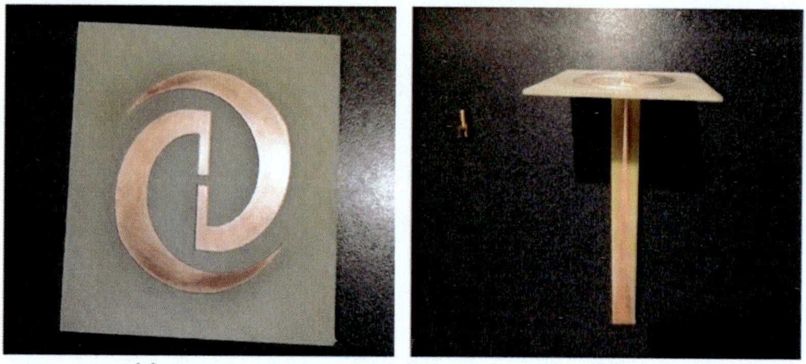

(a) (b)

Fig. 4. Manufacturing of LSA (a) front view, (b) cross-sectional view.

2.2 Logarithmic Periodic Antenna (LPA) Design

- **Initial frequency parameters**
  Frequency range (Fu): 2.4 GHz.
  Scale factor (τ): 0.802.
  Spacing factor (fl): 0.13 mm.
- **Number of dipoles**
  The number of elements in the array is given by the following equation.

$$Log(Fu) - Log(Fl) = (n-1)Log\left(\frac{1}{\tau}\right) \tag{6}$$

where:
Fu = 2.4 GHz
Fl = 500 MHz
τ = 0,802

$$Log\left(2.4 * 10^9\right) - Log\left(500 * 10^6\right) = (n-1)Log\left(\frac{1}{0.802}\right)$$

Then,

Number of elements $n = 7$.
Next, the dipole lengths are calculated.

$$Ln = \frac{\mathrm{c}}{(2 * fl)} = 50 \text{ mm} \tag{7}$$

- **Dipole measurements**

Table 4. Dipole measurements

N	L (length)	L (mm)	R (spacing)	R (mm)	Ln (L)-8/3	W (mm)
L1	0,08415975	84,16	0,008865248	8,87	1,766050	18,37
L2	0,07279818	72,80	0,00766844	7,67	1,621024	15,89
L3	0,06297042	62,97	0,0066332	6,63	1,475998	13,75
L4	0,05446942	54,47	0,005737718	5,74	1,330972	11,89
L5	0,04711604	47,12	0,004963126	4,96	1,185947	10,28
L6	0,04075538	40,76	0,004293104	4,29	1,040921	8,90
L7	0,03525340	35,25	0,003713535	3,71	0,895895	7,70

- **Calculation of dipole distances**

$$\tau = \frac{Dn}{Dn + 1} = 0,802 \tag{8}$$

where:
$Dn = Dis \tan ce\ of\ dipole$
1. $Dn = 8,\ 87$
2. $Dn + 1 = \tau\mathrm{x(Dn)} = 7,\ 64\,\mathrm{mm}$
3. $Dn + 2 = \tau\mathrm{x(Dn + 1)} = 10,\ 93\,\mathrm{mm}$
4. $Dn + 3 = \tau\mathrm{x(Dn + 2)} = 8,\ 76\,\mathrm{mm}$
5. $Dn + 4 = \tau\mathrm{x(Dn + 3)} = 7,\ 033\,\mathrm{mm}$
6. $Dn + 5 = \tau\mathrm{x(Dn + 4)} = 5,\ 64\,\mathrm{mm}$
7. $Dn + 6 = \tau\mathrm{x(Dn + 5)} = 4,\ 52\,\mathrm{mm}$

- **Simulation of the Logarithmic Periodic Antenna**
  The simulated structure for the periodic antenna is shown in Fig. 5, its main feature is the 7 dipoles, which are designed on the front and front side of the plate; this model was made to obtain greater resonance efficiency in the 2.4 GHz band.

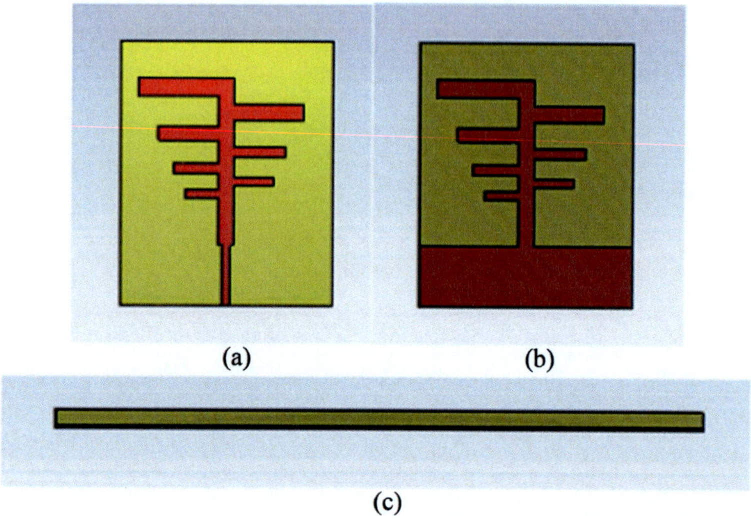

Fig. 5. Simulation of LPA. (a) front view, (b) back view, (c) cross-sectional view.

Figure 6 corresponds to parameter S11 of the Logarithmic Periodic Antenna. This antenna is designed to operate in a frequency range of 2.4 GHz.
The S parameter results in an operating frequency of 2.453 GHz at −13 dB.

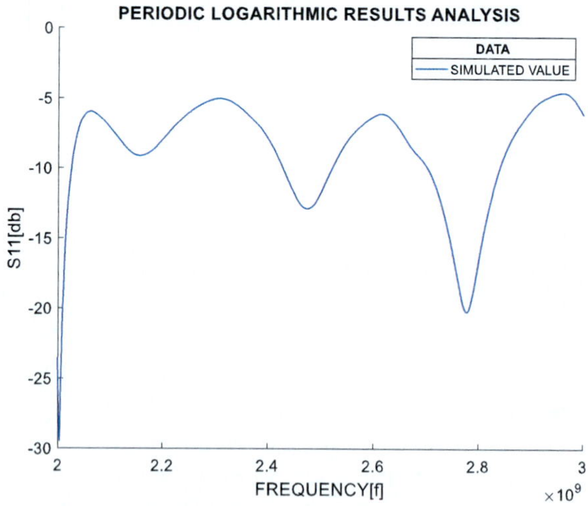

Fig. 6. Parameters S11 of LPA.

- **Manufacturing of the Logarithmic Periodic Antenna**

  The logarithmic periodic antenna, whose prototype is shown at the Fig. 7, corresponds to a double-sided antenna terminated in a horizontal line in order to improve the frequency range at which it works.

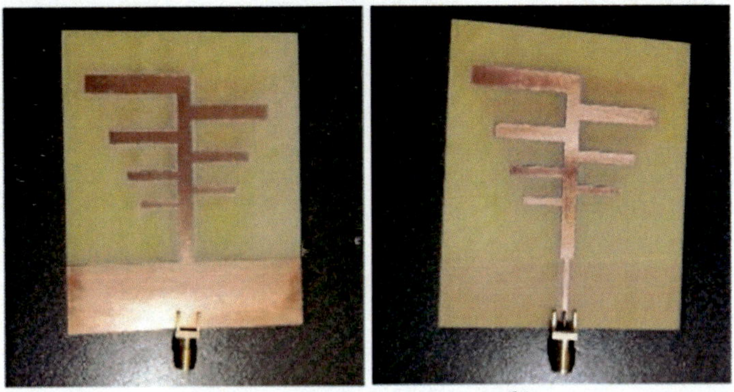

(a) (b)

Fig. 7. Manufacturing of LPA (a) front view, (b) rear view.

2.3 Voltage Multiplier Circuit Design

A Voltage Multiplier is an electrical circuit that generates a high voltage DC direct current from a low voltage alternating current source Cao pulsing to a higher voltage direct current source by means of diode and capacitor stages [12].

The circuit board consists of two stages: the voltage multiplier and the output control.

The circuit of the system simulated in Proteus is shown in Fig. 8, with all its elements which are: 4 Schottky diodes (BAT 43) and 2 rectifier diodes 1n4007, 4 ceramic capacitors of 1 nF and 2 connectors of 2 ways.

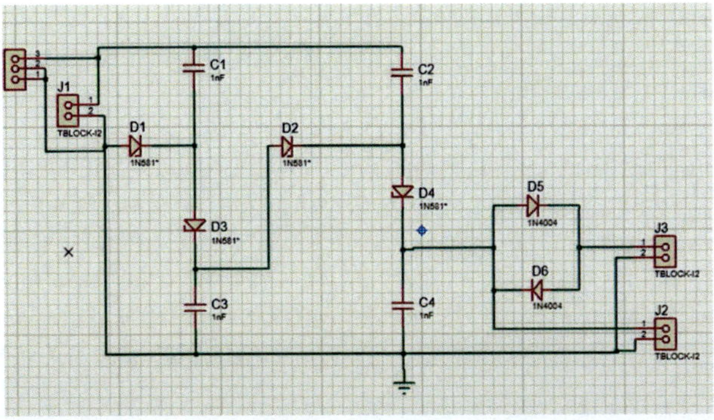

Fig. 8. Simulation of voltage multiplier circuit.

- **Manufacture of the voltage multiplier circuit**

Once the simulation was done, we proceeded to make the PCB board using Proteus software sketched in Fig. 9, to select the width of the conductor planes, size of the board and position of the elements.

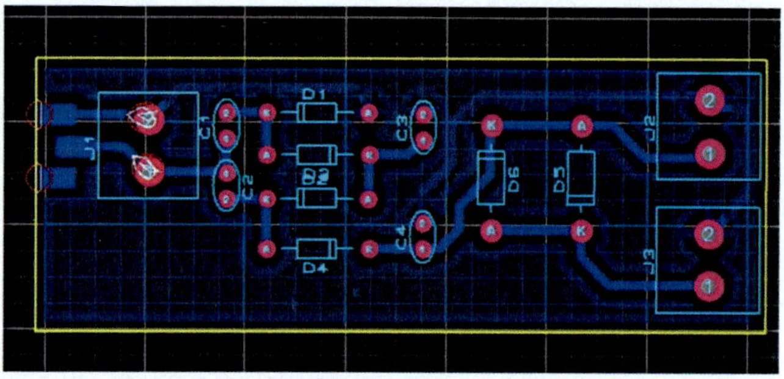

Fig. 9. PCB board of voltage multiplier circuit.

After the fabrication of the board, the electronic elements of the circuit were soldered. The Fig. 10, shows the implemented circuit.

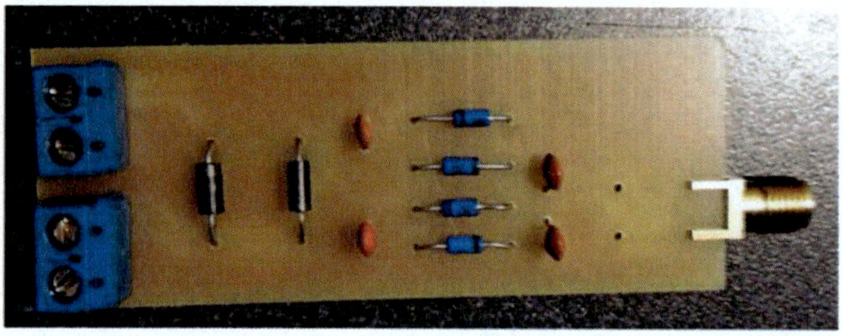

Fig. 10. Implementation of voltage multiplier circuit.

3 Discussion of Results

Each prototype of the proposed antennas has been designed, simulated and measured. The design and simulation were performed using CST Studio software. Printing as a practical process to perform the actual measurements of the antennas was performed on fr4 with copper cladding. Once the antennas were printed, the performance tests were performed using the MiniVNA-TINY antenna analyzer which is a device that measures frequencies from 1 MHz to 3 GHz, a computer, USB cable and the miniVNA-TINY device software.

The antenna design, simulation discussion and test results are explained below.

3.1 Electromagnetic Energy Harvesting System Performance Tests

- **System testing without storage**

The storage tests were performed in different environments, due to the presence of different electromagnetic fields, in order to obtain real system data without using a storage system.

Table 5, shows the results of the tests in each environment.

Table 5. Tests results in some environments

Environment	LSA	LPA
Internet modem	19 mV	16 mV
Internet router	25 mV	19 mV
Terrace with offset antennas	33 mV	25 mV

- **System tests with rechargeable storage battery**

For each storage test, the battery voltage (9V rechargeable battery) was determined before and after the energy harvesting process. The tests were performed by discharging the battery approximately halfway throughits storage, and measurements were taken after 30 min, 60 min and 120 min respectively; once the time had elapsed in each environment the following results were obtained. Table 6 shows the Charging times with LSA and Table 7 shows the Charging times with LPA.

Table 6. Charging times with LSA

Environment	Charging time					
	0 min–30 min		31 min–60 min		61 min–120 min	
	Vo (mV)	Vf (mV)	Vo (mV)	Vf (mV)	Vo (mV)	Vf (mV)
Internet modem	4500	4507	4507	4522	4522	4547
Internet router	4500	4505	4505	4511	4511	4523
Terrace with offset antennas	4500	4513	4513	4534	4534	4569

Table 7. Charging times with logarithmic periodic antenna

Environment	Charging time					
	0 min–30 min		31 min–60 min		61 min–120 min	
	Vo (mV)	Vf (mV)	Vo (mV)	Vf (mV)	Vo (mV)	Vf (mV)
Internet modem	4500	4509	4509	4519	4519	4527
Internet router	4500	4512	4512	4522	4522	4537
Terrace with offset antennas	4500	4516	4516	4546	4546	4586

- **Storage graphs in the different loading environments**

Figure 11 shows the battery storage obtained by the Logarithmic Spiral Antenna in the 3 different environments in which the loading was performed: internet modem 47 mV, Internet router 23 mV and Terrace with offset antennas 69 mV.

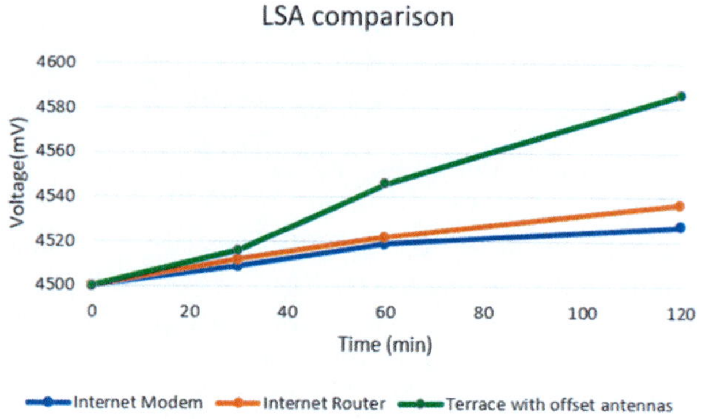

Fig. 11. LSA storage comparison.

Figure 12 shows the battery storage obtained by the Logarithmic Periodic Antenna in the 3 different environments in which the loading was performed: internet modem 27 mV, Internet router 37 mV and Terrace with offset antennas 83 mV.

3.2 Frequency Tests Using the MiniVNA-Tiny Spectrum Analyzer

The miniVNA Tiny spectrum analyzer depicted in Fig. 13 was used to test the effectiveness of the antenna design.

The Mini VNA-TINY is a very compact antenna analyzer with USB connection. It has a wide frequency range from 1 to 300 MHz. In addition, it is a network analyzer that not only allows SWR measurement. As a two-port system it can be used for transmission measurements of band filters or amplifiers. The device functions as a vector analyzer suitable for typical quadrupole measurements of S11 parameters.

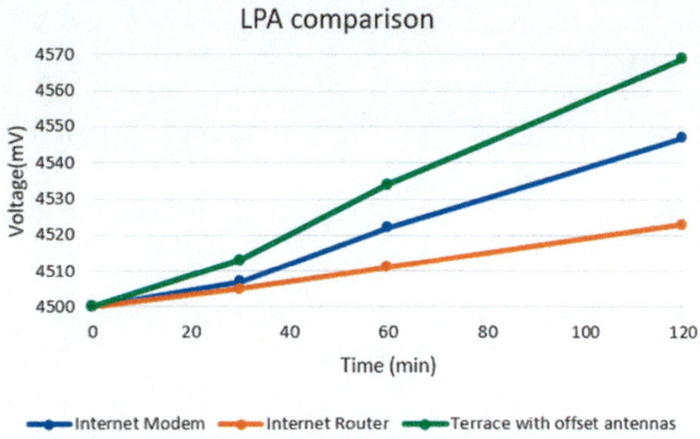

Fig. 12. LPA storage comparison.

This software is available for Windows, MAC and Linux, a new Android application is also available [13, 14].

This device is connected via USB cable to the computer and runs its software.

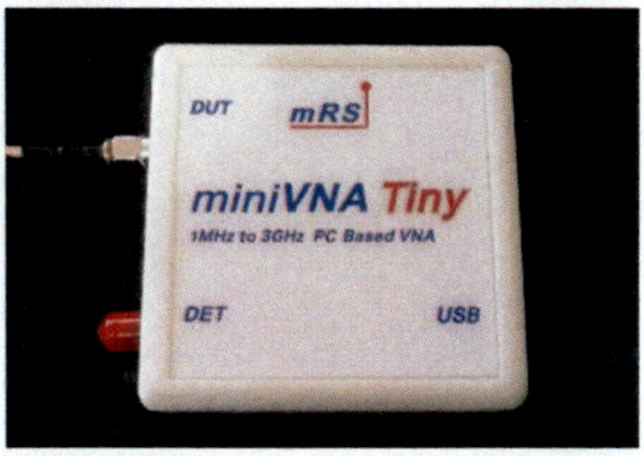

Fig. 13. miniVNA Tiny spectrum analyzer.

3.3 Comparison Between Simulated and Measured Value

- **Logarithmic Spiral Antenna analysis**

The simulated and measured reflection coefficient of the LSA is shown in Fig. 14. It is mainly compared between the simulated and measured values. The values obtained in the measurement with respect to the simulated one, operate optimally as they work at the desired frequency at 2.4 GHz, there is a small variation with respect to the desired frequency is due to losses in the material, solder and connector.

The simulator value of up to −8 dB of the antenna is optimal to implement.

The following equation is used to calculate the error between the simulation values and the actual measurement, given by the following equation:

$$\%error = \frac{|V_{aprox} - V_{real}|}{V_{real}} * 100\% \quad (12)$$

In the simulation the percentage error with respect to the desired resonance frequency (2.4 GHz) is:

$$\%error = \frac{|2.48 - 2.4\,GHz|}{2.4\,GHz} * 100\% = 3.33\%$$

In the measured value the error with respect to the resonance frequency (2.4 GHz) is:

$$\%error = \frac{|2.54 - 2.4\,GHz|}{2.4\,GHz} * 100\% = 5.83\%$$

From the comparison, it is clear that its error rate is minimal in real, making it an antenna that operates normally at 2.4 GHz which is the desired working frequency.

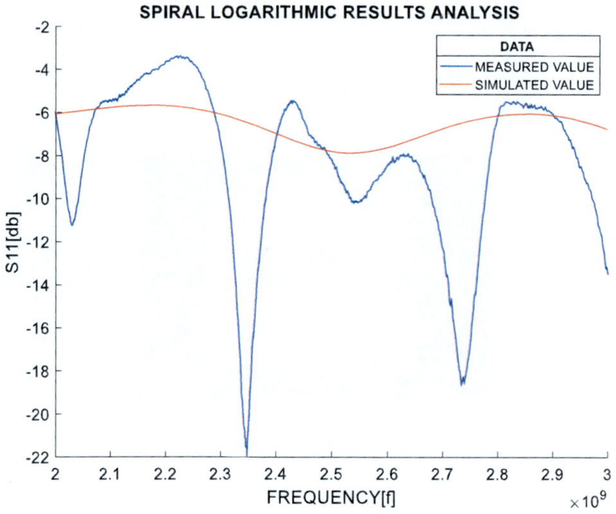

Fig. 14. LSA result analysis.

Analyzing the center frequency, the simulator excludes the secondary frequencies within 2.35 GHz and 2.75 GHz respectively, however, there is a bandwidth of about 400 MHz generated by the union of both segments: the antenna and the adapter as such.

Logarithmic Periodic Antenna Analysis

For the LPA, the simulation of the reflection coefficient and its respective comparison between the measured values, can be seen in Fig. 15 and the waveform obtained with the real data, operates normally in a frequency range 2.4 GHz with values of −13 dB. By designing and implementing this antenna is obtained in the real part multiple operating frequencies that make it a high-end antenna for a specific purpose in this case for the collection of electromagnetic energy.

To calculate the error has been considered at a frequency of 2.39 GHz.

In the simulation the percentage error with respect to the desired resonance frequency 2.4 GHz is:

$$\%error = \frac{|2.39 - 2.4\,GHz|}{2.4\,GHz} * 100\% = 0.42\%$$

In the measured value the error with respect to the resonance frequency (2.4 GHz) is:

$$\%error = \frac{|2.45 - 2.4\,GHz|}{2.4\,GHz} * 100\% = 2.08\%$$

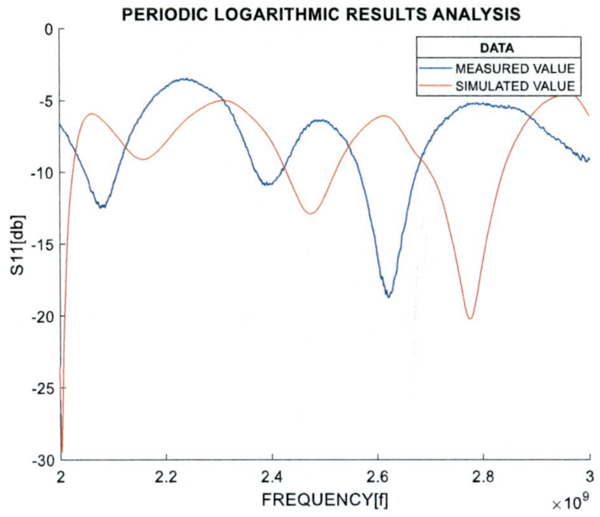

Fig. 15. LPA result analysis.

In the simulation a defined resonance of 2.48 GHz is obtained, likewise in the measurement it provides multiple resonances that operate in the frequency range of 2.4 and 2.6 GHz, this resonance is due to the material and its welds with the 50 Ohms SMA connector and the losses that occur in the manufacture; therefore, the antenna in mention is suitable for electromagnetic energy collection, especially in the range of 2.4 to 2.6 GHz whose margin of error is 2.08%.

4 Conclusions

In this research, an electromagnetic energy collection system was implemented using two types of logarithmic patch antennas: spiral and periodic, both designed to work at a central frequency of 2.4 GHz. The energy collected by the antennas is small, so a voltage amplifier circuit is used to rectify and filter the output signal for storage.

The storage in rechargeable batteries of the spiral logarithmic antenna is exponential in the 3 environments to which the system was subjected for charging, resulting in the environment near an Internet modem a total load of 47 mV, near the Internet router the final load was 23 mV and on the terrace of the FISEI where there are offset antennas the final load was 69 mV after 120 min of charging.

The storage in rechargeable batteries of the periodic logarithmic antenna, as in the spiral logarithmic antenna, is exponential in the 3 environments to which the system was subjected for charging, resulting in the environment near an Internet modem a total load of 27 mV, near the Internet router the final load was 37 mV and on the terrace of the FISEI where there are offset antennas the final load was 86 mV after 120 min of charging.

Acknowledgements. The authors thank the Technical University of Ambato and the "Dirección de Investigación y Desarrollo" (DIDE) for their support in carrying out this research, in the execution of the project "Sistema de Captación de Energía Electromagnética para Abastecimiento de Energía en Terminales de Internet de las Cosas (IoT) en entornos de Quinta Generación (5G).", project code: SFFISEI 04.

References

1. Di Paolo Emilio, M.: Microelectronic Circuit Design for Energy Harvesting Systems. Springer International Publishing, Pescara-Italy (2017)
2. Adhikari, S., Inman, D.J.: Piezoelectric Energy Harvesting, 1st edn. Wiley, USA (2011)
3. Alsharif, M.H., Kim, S., Kuruoglu, N.: Energy harvesting techniques for wireless sensor. Symmetry **7**(11), 865–879 (2019)
4. Cerquera, B, Blanco, D.: Design and Simulation of a Rectenna to Harvest Electromagnetic Energy at 2.4 GHz. Colombia Catholic University, Bogota D.C (2020)
5. Panatik, K, Kamardin, K, Shariff, S.: Energy harvesting in wireless sensor networks: A survey. In: IEEE 3rd International Symposium on Telecommunication Technologies (ISTT), pp. 53–58 (2016)
6. Martínez, A.: Antennas for energy harvesting applications in the uhf band. Polytechnic University of Valencia, Spain (2014)
7. El país: The wifi signal, converted into electricity, https://elpais.com/elpais/2019/01/27/ciencia/1548603829_841509.html#:~:text=Ahora%2C%20investigadores%20de%20universidades%20de,ondas%20electromagn%C3%A9ticas%20en%20corriente%20continua. last accessed 25 June 2021
8. Sleebi, K.D., Deepti, D., Nasimuddin: RF energy harvesting systems: An overview and design issues. The international journal od RF and Microwave Computer-Aided Engineering Journal **29** (2018)
9. Mrnka, M., Vasina, P., Kufa, M., Hebelka, V., Raida, Z.: International Journal of Antennas and Propagation, 11–22 (2016)
10. Snehal, P., Sonal, G.: Design and implementation of microstrip antenna for RF. Energy **10**, 487–490 (2017)
11. Electrodrone: visited (01 June 2021). Contact: Info@electrodrone.net. https://electrodrone.net/
12. Johnson, R.: Antenna Engineering Handbook, New York. McGraw-Hill, EEUU (1997)
13. Palomino Vera, K.: Prototype generator of electric energy by the use of an elliptical bicycle for the illumination of a sports environment. Technological University of Peru, Lima (2016)

14. Int.wimo: page. https://int.wimo.com/en/minivna-tiny. last accessed 03 July 2021
15. Ismail, N., Ghani, R.A.: Advance devices using piezoelectric harvesting energy. IEEE Student Conference on Research and Developement, Putrajaya, Malaysia (2013)

Author Index

M. Botto-Tobar et al. (Eds.): *Latest Advances in Electrical Engineering, and Electronics*, LNEE 933, pp. 251–252, 2022.
https://doi.org/10.1007/978-3-031-08942-8